Chickpea (Gram):
Status and Cultivation Technology

Dr. S.S. Shekhawat (born. May 15, 1961), Sr. Scientist, AICRP on Chickpea ARS, Durgapura, Jaipur was awarded Ph.D. in Plant breeding and Genetics from RAU, S.K.N. College of Agriculture, Jobner, Jaipur (Raj) in 2002. He is working on crop improvement since 1990 and working in legumes from 1996. He has contributed in seven varieties and several donors taking care, in view of different climatic conditions, biotic and abiotic stresses, consumer preference , value addition and export quality, more than fifteen research papers, number of book chapters, abstracts, popular articles , folders and pamphlets.

Chickpea (Gram): Status and Cultivation Technology

Dr. S.S. Shekhawat

Senior Scientist
All India Coordinated Project on Chickpea
{Swami Keshwanand Rajasthan Agricultural University,
Bikaner-334002 (Rajasthan)}
Agricultural Research Station
Durgapura, Jaipur, Rajasthan - 302 018

SCIENTIFIC
PUBLISHERS (INDIA)

Published by:

Scientific Publishers (India)
5 A, New Pali Road, P.O. Box 91
Jodhpur 342 001 (India)

Branch Office
Scientific Publishers (India)
4806/24, Ansari Road, Daryaganj
New Delhi - 110 002 (India)

E-mail: info@scientificpub.com
Website: www.scientificpub.com

Print : 2022

ISBN: 978-81-7233-851-0

Typeset by Rajesh Ojha
Printed in India

This Manuscript

CHICKPEA (GRAM) Status and Cultivation Technology

is

Dedicated to my respected father

Late Thakur Raghuveer Singh Ji Shekhawat

FOREWORD

Chickpea occupies a unique place in food legumes in India agriculture and is best option for diversification and intensification because it fixes atmospheric nitrogen to the soil and it has ability to thrive well in poor and marginal lands. It's intrinsic value in terms of higher protein plays an important role in food, feed, soil health, industry, eco-system and sustainable crop production. It provides nutritional security to millions of people suffering from protein malnutrition as it is rich source of protein, minerals, vitamins, sugar, fat, crude fiber, dietary fibre and several amino acids required for human health. Importance of chickpea for improving nutritional security in developing countries is well known. Chickpea becomes more important in our country where majority of people are poor and vegetarian, depends on low priced food for their dietary requirement. It has ability to tolerance drought and heat compared to other seasonal food legumes. Traditionally, it is cultivated as a rain-fed crop and sometimes as a irrigated one. Chickpea has wide adaptability as it is cultivated from the equator to 50ºN. It has meditational value and helpful to overcome so many diseases. With the unique ability of biological nitrogen fixation, low water requirement and deep root system, legumes thrive well under adverse weather condition and capacity to withstand in drought condition. It has become an important component of dry land farming system across the globe.

The supply of quality seed of improved varieties is important not only for realizing full potential of varieties but for retaining faith of farmers in high yielding technology. It is the key input for enhancing the agricultural production and increasing the food availability to feed the growing population of India and the world as well. Neither the poor quality seed of superior variety nor the high quality seed of inferior variety can serve the purpose of crop production. Seed account for a small part of the total cultivation expenditure. Without good quality seed, the investment on fertilizer, irrigation, pesticides and other inputs will go in vain. The need of production technology is well recognized for increasing the yield. In any cropping system, production technology is required for increasing yield, resistance against biotic and abiotic stress, value addition. This publication **"Chickpea (Gram): Status and Cultivation Technology"** provides latest information about constraints and strategies in chickpea cultivation, chickpea breeding, physiology of drought tolerance in chickpea, production technology in chickpea, integrated nutrient management in chickpea, use of bio-fertilizer in chickpea, integrated weed management in chickpea, integrated disease management in chickpea, integrated insect-pest management in chickpea, integrated nematode management in chickpea, quality seed production in chickpea and safe storage etc. which will be helpful to the researchers, students, extension workers, professionals and the last user of the technology **"The farmer".**

PREFACE

Pulses are the back bone of the national economy, nutritional security of vegetarian society, soil health and sustainability of environment. In India, we are facing the problem of poor fertility of soil, lack of irrigation, poor condition of farmer and pre-dominating vegetarian mass, and in the present scenario pulses become more important because, only pulses opens the door of sustainability in the field of agriculture. Being a rich source of protein, minerals, vitamins, crude fiber and several amino acids essential for human health, pulses are considered as a health food and offer nutritional security to millions of people suffering from protein malnutrition. However, it fixes atmospheric nitrogen in the soil.

The important pulses cultivated in India are Chickpea, lentil, fieldpea, rajmah, lathyrus, pigeonpea, mungbean, urdbean, cowpea, mothbean and horsegram. A cross the globe, Chickpea is an important pulse cultivated in an around 40 countries. It ranks first in area and production at national level. Present manuscript "**Chickpea (Gram): Status and Cultivation Technology**" includes, The opening chapter gives an overview on chickpea research in india, Status and Cultivation Technology, Introduction about chickpea, constraints and strategies in chickpea cultivation, Subsequent chapters discuss chickpea breeding, chapters on physiology of drought tolerance in chickpea, production technology in chickpea, nutrient management in chickpea, role of bio-fertilizers in chickpea, integrated weed management in chickpea provide comprehensive information on production management of the crop, chapters on management of stress, diseases, insect-pest, nematodes deals with different aspects of abiotic and biotic stresses, their efficient management in integrated manner. The last two chapters deal with seed production and post harvest technologies in perspective of safe storage. We hope this compendium will serve as a standard reference book on chickpea and encourage further research in developing high yielding, effective technologies, value addition, consumer preference and export quality to earn a handsome forex which leads to strong economic contribution to our nation. It will be helpful to the researchers, students, extension workers, professionals and the farmers.

The Author feels deep gratitude towards **Dr. N.P. Singh, Project Coordinator, AICRP on chickpea, Indian Institute of Pulse Research, IIPR, Kanpur,** UP, Prof. Swaroop Singh, Zonal Director Research, ARS, Durgapura, for their consistent encouragement and guidance in bringing out this publication. I also want to thank Dr. S.J. Singh, Senior Chickpea Breeder, Dr. S.C. Gupta, Senior Chickpea Physiologist, Dr. O.P. Khedar, Senior Pulse Breeder, Dr. O.P. Sharma, Senior Chickpea Pathologist; Dr. M.M. Singh, Senior Chickpea Agronomist, Dr. Surendra Singh, Senior Agronomist, Dr. Yogendra Singh, Incharge Cum Senior Millet Breeder, Dr. Hoshiyar Singh, Senior Wheat Breeder, Dr. Mahesh Srimali, (HOD) Plant Breeding and Genetic Section, Incharge cum senior wheat breeder, Dr. R.S. Sain, Senior Plant Breeder, Dr. S.M. Yadav, Senior Chickpea Nematologist, Dr. Kuldeep Sharma, Senior Pulse Pathologist, Dr. V.D. Fageria, Senior Pulse Agronomist, Dr. S.S. Bareth, Senior Entomologist, Dr. R.S. Mahla, Senior Pulse Entomologist, for taking keen interst and giving their valuable advice and kind support, Prof. Govind Singh Ji (Chui) (Jai Narayan Vyas University,

Jodhpur), Dr. Mahipal Singh Rathore, Dr. Jaipal Singh Rathore, Shri Suresh Kumar Sharma (P.A. AICRP on Millet) for their valuable guidance, constructive suggestion keen and sustained interest, incessant encouragement, propitious assistance, kind and gracious patronage in preparation of this manuscript. It is my feeling that my late Father's soul had been a great source of inspiration to me all along my task accomplished. Author is grateful to all the contributors from S.K. Rajasthan Agricultural University, Bikaner and Agricultural Research Station, Durgapura, Jaipur to share their experiences on chickpea research. Author is also thankful to all family members for their moral support at every step.

Jaipur **S.S. Shekhawat**

LIST OF TABLES

LIST OF FIGURES

CONRTIBUTORS

Dr. A.S. Baloda
Sr. Scientist (Entomology)
SK RAU Agricultural
Research Station,
Durgapura-302018, Jaipur

Dr. A.S. Jat
SMS (Agronomy)
Krishi Vigyan Kendra,
Badun, Ujjni, SDVBPUAT,
Meerut (UP)

Dr. Baldev Ram
Scientist (Agronomy)
MPUAT, Agricultural
Research Station, Ummed-
ganj, Kota-324001, Raj.
baldev.ram@gmail.com

Dr. B.S. Shekhawat
Senior Scientist (Agronomy)
SK RAU, Agricultural
Research Station,
Durgapura-302018, Jaipur

Dr. G.L. Yadav
Senior Scientist (Agronomy)
SK RAU, Agricultural
Research Station,
Durgapura-302018, Jaipur

Dr. Hoshiyar Singh
Senior Scientist
(Plant Breeding)
SK RAU, Agricultural
Research Station,
Durgapura-302018, Jaipur
hoshiar.singh@excelcropcare.com

Jagdish Prasad
Scientist (Plant Pathology)
SK RAU, Agricultural
Research Station,
Durgapura-302018, Jaipur

Dr. Kuldeep Sharma
Senior Scientist
(Plant Pathology)
SK RAU, Agricultural
Research Station,
Durgapura-302018, Jaipur
kuldeeppatho@gmail.com

Dr. L.D. Sharma
Senior Scientist
(Plant Breeding)
SK RAU, Agricultural
Research Station,
Durgapura-302018, Jaipur

Dr. M.M. Singh
Sr. Scientist (Agronomy)
SK RAU Agricultural
Research Station,
Durgapura-302018, Jaipur

Dr. N.P. Singh
Project Coordinator, AICRP
on chickpea, Indian Institute
of Pulse Research, IIPR,
Kanpur, U.P., India

Dr. O.P. Khedar
Pulse Breeder Cum Incharge
AICRP on MULLaRP
SK RAU, Agricultural
Research Station,
Durgapura-302018, Jaipur
khedarop@gmail.com
khedarop@hotmail.com

Dr. O.P. Sharma
Senior Scientist
(Plant Pathology)
SK RAU, Agricultural
Research Station,
Durgapura-302018, Jaipur

Dr. Pradeep Singh
Assistant Professor (Plant
Pathology), SK RAU,
Agricultural Research
Station,
Durgapura-302018, Jaipur

Dr. R.S. Mahla
Senior Scientist (Entomology)
SK RAU, Agricultural
Research Station,
Durgapura-302018, Jaipur

Dr. R.S. Sain
Senior Scientist (STR)
SK RAU, Agricultural
Research Station,
Durgapura-302018, Jaipur

Dr. R.V. Singh
Groundnut Breeder Cum
Incharge AICRP on
Groundnut
SK RAU, Agricultural
Research Station,
Durgapura-302018, Jaipur
singhrvgnut@gmail.com

Dr. S.C. Gupta
Senior Scientist
(Plant Physiology)
SK RAU, Agricultural
Research Station,
Durgapura-302018, Jaipur

Dr. S.M. Yadav
Senior Scientist Cum
Incharge Nematology, SK
RAU, Agricultural Research
Station,
Durgapura-302018, Jaipur

Dr. Surendra Singh
Senior Scientist (Agronomy)
SK RAU, Agricultural
Research Station,
Durgapura-302018, Jaipur
singhmanohar.2012@gmail.com

Dr. S.R. Dhaka
Senior Scientist (Entomology)
Cum Officer Incharge
SK RAU, Agricultural
Research Sub Station,
Diggi-302018, Tonk

Dr. S.S. Bareth
Senior Scientist (Entomology)
SK RAU, Agricultural
Research Station,
Durgapura-302018,
Jaipur

Dr. S.S. Punia
Scientist (Plant Breeding)
Cum Incharge MULLaRP
MPUAT, Agricultural
Research Station,
Kota, Raj.
sumerspunia@hotmail.com

Dr. S.S. Shekhawat
Senior Scientist
(Plant Breeding) SK RAU,
Agricultural Research
Station, Durgapura-302018,
Jaipur
s.s.shekhawat61@gmail.com

Dr. V.D. Fageria
Senior Scientist (Agronomy)
SK RAU, Agricultural
Research Station,
Durgapura-302018,
Jaipur

Dr. Yogendra Singh
Senior scientist (Plant
Breeding) Cum Incharge
AICRP on millet
SK RAU, Agricultural
Research Station, Durgapura-
302018, Jaipur
Yogendrasingh448@gmail.com

TEXT ABBREVIATIONS

@	At the rate of
AAU	Assam Agriculture University
ai/ha	Active ingredient per hectare
ANGRAU	Acharya N.G. Ranga Agriculture University, Hyderabad
AP	Andhra Pradesh
APAU	Andhra Pradesh Agriculture University
ARS	Agriculture Research Station
Avg.	Average
BARC	Bhabha Atomic Research Center
BAU	Bihar Agriculture University
BCMV	Bean common mosaic virus
BGM	Botrytis grey molds
BHU	Banaras Hindu University
BLB	Bacterial leaf blight
BNF	Biological nitrogen fixation
BOAA	Beta N-oxalyl-L-alpha-beta diaminopropionic acid
BSP	Breeder seed production
Ca	Calcium
Carb.	Carbohydrate
CCS HAU	Choudhary Charan Singh Haryana Agricultural University
CG	Chhattisgarh
cms	Centimeters
CMMV	Cowpea mild mottle virus
CSAU	Chandra Shekhar Azad Agricultural University, Kanpur (UP)
CSC	Central Seed Committee
CVRC	Central Varietal Release Committee

CZ	Central zone (Gujarat, Maharashtra, M.P., Chhattisgarh and parts of Raj.)
DAE	Days after emergence
DAS	Days after sowing
DUS	Distinctness, uniformity and stability
ELISA	Enzyme linked immunosorbent assay
ELS	Early leaf spot
ETL	Economic threshold level
EXIM	Export import
FAO	Food and Agriculture Organization
Fe	Iron
FIR	Fungicide, Insecticide, *Rhizobium*
FYM	Farm yard manure
GBPUA&T	G.B. Pant University of Agriculture & Technology, Pant Nagar (Uttra Khand)
g	Gram
g/lit.	Gram per litre
g/kg	Gram per kilogram
GOI	Government of India
GP	Germplasm
Guj.	Gujarat
ha	Hectare
ha^{-1}	Per hectare
Har.	Haryana
HP	Himachal Pradesh
IARI	Indian Agricultural Research Institute, New Delhi
ICAR	Indian Council of Agricultural Research
ICRISAT	International Crop Research Institute for Semi Arid Tropics, Hyderabad
IDM	Integrated disease management
IET	Initial Evaluation Trial
IGKVV	Indira Gandhi Krishi Vishva Vidyalaya, Raipur (CG)
IIPR	Indian Institute of Pulses Research, Kanpur (UP)
IMSCS	Indian Minimum Seed Certification Standards
IPM	Integrated pest management
IVT	Initial Varietal Trial
J & K	Jammu and Kashmir
JNKVV	Jawahar Lal Nehru Krishi Vishva Vidyalaya, Jabalpur (MP)
K cal	Kilo-calories
kg	Kilogram
kg/ha	Kilogram per hectare
kg/ha/year	Kilogram per hectare per year
lit/ha	Liter per hectare
m	Meter
MS	Maharashtra

mg	Milligram
m ha	Million hectare
ml/kg	Milliliter per kilogram
mm	Millimeter
Mod.	Moderately
MP	Madhya Pradesh
MPKV	Mahatma Phule Krishi Vidyapeeth, Rahuri, (MS)
MS	Maharashtra state
MSCS	Minimum Seed Certification Standards
m tonnes	Million tones
m t	Million tones
NBDC	National Biofertilizer Development Centre
NBPGR	National Bureau of Plant Genetic Resources
NDUA&T	Narendra Dev University of Agriculture and Technology, Faizabad (UP)
NEPZ	North-east plane zone (Bihar, Jharkhand, West Bengal and Eastern UP)
NEZ	North-east zone
NHZ	North hill zone
NSC	National Seeds Corporation
NWPZ	North-west plane zone (Haryana, Punjab, Rajasthan, Western UP)
NWZ	North-west zone
OUAT	Orissa University of Agriculture and Technology
P	Phosphorus
/gm	Per gram
PAU	Punjab Agriculture University, Ludhiana (Punjab)
PGPR	Plant Growth Promoting *Rhizobecteria*
P^H	Hydrogen ion concentration
PKV	Punjab Rao Deshmukh Krishi Vidyapeeth, Akola (MS)
PORS	Puddukutai Operational Research Station, Barhampur (WB)
PSB	Phosphate solublizing bacteria
q/ha	Quintal per hectare
Raj.	Rajasthan
Res.	Resistant
RH	Relative humidity
RR	Root rot
SAU	State Agricultural University
Sel.	Selection
SFCI	State Farms Corporation of India
SKRAU	Swami Keshwanand Rajas-than Agricultural Univer-sity, Bikaner
SOC	Soil organic carbon

SOM	Soil organic matter
Sp.	Species
SPS	Single plant selection
SRR	Seed replacement rate
SSC	State Seeds Corporation
ST	Seed treatment
Sui.	Suitable
Sus.	Susceptible
SVBPUAT	Sardar Vallabh Bhai Patel University of Agriculture & Technology, Meerut (UP)
SZ	South zone (Andhra Pradesh, Tamil Nadu, Orissa, Karnataka)
TN	Tamil Nadu
TNAU	Tamil Nadu Agriculture University, Coimbatore (TN)
UP	Uttar Pradesh
USA	United States of America
var.	Variety
viz.	Namely
VPKAS	Vivekanand Parvtiya Krishi Anusandhan Sansthan, Almora (Uttranchal)
WANA	West Asia and North Africa covering Turkey in North to Ethopia in the South and Pakistan in East to Morocco in West
WB	West Bangal
YMV	Yellow mosaic virus

CONTENTS

1

Chickpea (Gram) Status and Cultivation Technology: An Overview

S.S. Shekhawat

INTRODUCTION

Globally, grain legumes are second most important group of crops after cereals in area and production. The important legumes are Soyabean, groundnut, dry beans, dry peas, chickpea, lentil and pigeonpea. Together with cereals, they are dominant staple for human food and animal feed. About one third of dietary protein and almost an equal amount of processed vegetable oils estimated to come from grain legumes. Though, India is the largest producer (25%) of pulses in the world, both in quality and quantity. It restores soil fertility and has unique position in cropping system and sustainable agriculture, products.

Importance of pulses for improving the household nutritional security in our country on account of giving balance amino acid composition in consumption with cereals. Whereas, at the same time we are importing 20% and consuming 27% share. Our current requirement of pulses is 17 m tonnes, whereas, the production is 15.19 m tonnes. The import of pulses has increased from 0.5 m tonnes (1995) to 1.87 m tonnes (2007). We are exporting chickpea, lentil and pigeonpea dal in very small quantities, however, production is not meeting up with the demand and for that reason we are importing pulses in a large quantity from different countries. Dry pea, pigeonpea, *kabuli* chickpea, and urdbean are the major imported pulses in India. Globally, the countries like Canada, Myanmar, China, France, Australia and USA are the important exporters, whereas, India, Spain, Egypt, Italy and Bangladesh are the main importers. Globally, chickpea (*Cicer arietinum* L.) is the fourth most important grain legume with production of 9.31 m tonnes from an area of 11.7 m ha and productivity of 796 kg/ha (FAO STAT 2007). Chickpea is an important cold season crop in the world and India is the second largest producer of global chickpea, after Mexico. In India chickpea contribute 40% of total pulse production. It is an important pulse crop grown during *rabi* season in tropics and *spring* in the temperate region of the world. The genus *Cicer* comprises 34 perennial species, 8 annual wild and one annual cultivated species. Most of these species are found in the West Asia and North Africa (WANA) region covering Turkey in the north, Ethiopia in the south and Pakistan in the east, Morocco in the west. Out of these species *Cicer arietinum* L. is the only cultivated specie. Two main categories of chickpeas (*desi* and *kabuli* types) are distinguished mainly by their seed characteristics. The *desi* types also known as Bengal gram and *kabuli* known as Garbamzos. Though, India is the largest producer of chickpea, however, generally India imports chickpea from Turkey, Australia, Canada, USA etc and major part is *kabuli* types (extra bold seeded). It is because of non-availability of extra bold seeded varieties. The adaptation and drought tolerance of *kabuli* types is also very poor. *Desi* chickpea is generally tolerant to drought and poor management.

Majority of Indian population is poor and vegetarian, depends on low priced food for meeting their dietary requirement, chickpea has become more important to mitigate the problem of protein energy malnutrition. Besides being rich source of highly digestible dietary protein (19.4%), it is also a rich source of calcium, iron, niacin, Vitamin B and Vitamin C. It also provides 396 K cal energy from 100 grams of seed. In general *desi* type is used as split *dal* or flour (*besan*) or sometimes sprouted seeds are also used to eat. The other preparations of *besan* are sweets, namkeen, snacks etc. Premature green seeds are also used as vegetable. *Kabuli* types are generally used as whole in the vegetable (mixed or alone). Natural selection has endowed the capacity to fix atmospheric nitrogen to the legumes and being legume, chickpea plays an important role in improving the fertility of the soil by nitrogen fixation with the help of *Rhizobium*. Under better management condition chickpea fixes up to 141 kg N/ha/ year. With the judicious use of *Rhizobium* culture, PSB and PGPR, legumes are able to meet much of their nutrient requirement. The ability to form this symbiosis reduces fertilizer costs to the farmers. It also improves the texture and structure of the soil. They help in mobilizing insoluble soil nutrients and bringing qualitative changes in physical properties of the soil. It is fact that if we want real sustainability, it is only possible by adding legumes in our farming system. Chickpea also provide nutritive and productive fodder to the cattle.

Chickpea crop is able to utilize the limited available moisture more efficiently through their deep root system and help substantially in loosening as well as improving physical properties of the soil. In general chickpea is more drought tolerant than cereals. The rapid increase in population in the past years has led to decline in per capita availability of pulses in the country. As per FAO/WHO recommendation the minimum pulse requirement is 108 g/capita in which chickpea share is 70 g/capita/day. The per capita availability of pulse is 42 g in Rajasthan while chickpea is 25 g/day in comparison to national average availability 18 g/day. Hence, there is an urgent need of boosting the chickpea production in the state as well as in country.

COMMON NAMES

Bengal gram, chana, gram, chhole, garbanzo bean, homes etc.

UNDESIRABLE NUTRITIONAL CONSTITUENT

Choline, oxalic acid, saponin, trypsin and chymotrypsin inhibitors.

ORIGIN

Cicer arietinum L. has its genesis from *C. reticulantum* in the middle-East which spread to Mediterranean Sea region, Ethiopia and Indian sub-continent where it has been cultivated since time immemorial.

GLOBAL SCENARIO

Chickpea (gram) is cultivated in over 42 countries across the globe. It is widely distributed in the world. Near about 42 countries grow chickpea but the countries like India, Pakistan, Turkey, Canada, Mexico, Iran, Australia, Ethiopia, Myanmar, Syria, Bangladesh, and Spain contribute 97% to the global production. It is premier pulse crop of India and is cultivated throughout the country except high altitudes and coastal region. MP, Rajasthan, UP, Maharashtra, AP and Karnataka are the major chickpea growing states contribute more than 90% of the total area (Table 1.1 & 1.2). Other chickpea growing states are Chhattisgarh, Gujarat, Haryana, Bihar, Orissa, Punjab, West Bengal, Tamil Nadu, Assam, Himachal Pradesh, J & K, Meghalaya, Nagaland, Tripura, Uttaranchal, Dadar & Nagar Haveli and Delhi.

Table 1.1. Area (m ha), production (m tonnes) and productivity (kg/ha) of Chickpea in India

Year	Area	Production	Productivity	Year	Area	Production	Productivity
1964-65	8.87	5.78	651	1988-89	6.81	5.13	753
1965-66	8.02	4.22	527	1989-90	6.47	4.22	652
1966-67	8.00	3.62	453	1990-91	7.52	5.36	712
1967-68	8.26	5.97	723	1991-92	5.58	4.12	739
1968-69	7.11	4.31	607	1992-93	6.45	4.42	684
1969-70	7.75	5.55	715	1993-94	6.35	4.98	783
1970-71	7.84	5.20	663	1994-95	7.54	6.44	853
1971-72	7.91	5.08	642	1995-96	7.12	4.98	700
1972-73	6.97	4.54	651	1996-97	6.85	5.57	813
1973-74	7.76	4.10	528	1997-98	7.56	6.13	811
1974-75	7.04	4.02	570	1998-99	8.47	6.80	803
1975-76	8.32	5.88	707	1999-00	6.15	5.12	833
1976-77	7.97	5.42	680	2000-01	5.19	3.86	744
1977-78	7.97	5.41	678	2001-02	6.42	5.47	853
1978-79	7.71	5.74	745	2002-03	5.91	4.24	717
1979-80	6.99	3.36	481	2003-04	7.05	5.72	811
1980-81	6.58	4.33	657	2004-05	6.71	5.47	815
1981-82	7.87	4.64	590	2005-06	6.93	6.00	808
1982-83	7.40	5.29	715	2006-07	7.63	6.33	830
1983-84	7.16	4.75	663	2007-08	7.58	5.75	759
1984-85	6.91	4.56	661	2008-09	7.97	7.06	885
1985-86	7.8	5.79	742	2009-10	8.21	7.48	895
1986-87	6.98	4.53	649	2010-11	9.21	8.22	896
1987-88	5.77	3.63	629	2011-12	9.01	7.58	841

Table 1.2. Area (m ha), production (m tonnes) and productivity (kg/ha) of major chickpea growing states

Year	Area	Production	Productivity	Area	Production	Productivity
	Madhya Pradesh			Maharashtra		
94-95	27.414	24.872	907	7.632	4.686	614

95-96	26.602	19.884	747	7.171	3.760	524
96-97	25.126	22.933	913	7.495	4.984	665
97-98	25.825	24.409	945	7.150	2.910	407
98-99	26.654	24.634	924	9.042	5.619	621
99-2K	25.719	25.351	986	9.320	6.000	644
2K-01	19.784	16.203	819	6.760	3.510	519
01-02	25.539	24.082	943	7.560	4.500	595
02-03	24.706	17.134	694	7.960	4.220	530
03-04	27.913	25.849	926	7.950	4.210	530
04-05	27.465	25.489	928	8.300	4.660	561
05-06	25.607	23.712	926	10.200	7.050	691
06-07	24.628	24.130	980	13.080	9.24	706
07-08	24.387	17.350	711	13.530	11.16	825
08-09	28.400	27.900	981	11.400	7.70	677
09-10	30.900	33.000	1071	12.900	11.10	863
		Rajasthan			**Andhra Pradesh**	
94-95	15.865	13.711	864	1.682	1.361	809
95-96	16.203	10.903	673	1.129	0.711	630
96-97	15.191	10.710	705	1.055	0.900	853
97-98	22.139	19.246	869	1.470	0.585	398
98-99	28.157	20.738	737	1.460	1.300	890
99-2K	9.753	6.779	695	1.630	0.951	583
2K-01	6.726	3.966	590	2.010	2.290	1139
01-02	9.696	7.355	759	2.850	3.630	1274
02-03	4.497	3.406	757	3.900	3.820	979
03-04	11.175	7.072	633	4.220	4.570	1083
04-05	10.352	7.730	747	3.410	3.450	1012
05-06	10.819	4.730	443	3.940	6.270	1591
06-07	10.10	8.720	863	6.020	6.530	1085
07-08	12.31	5.740	466	6.300	9.120	1448
08-09	12.60	9.800	779	6.100	8.600	1412
09-10	88.00	5.300	604	6.500	8.500	1308
10-11	16.42	13.86	844	-	-	-
11-12	14.34	9.91	691	-	-	-

	Uttar Pradesh			Karnataka		
94-95	10.365	9.480	915	4.412	2.216	503
95-96	10.059	6.983	694	2.934	1.416	485
96-97	9.166	8.524	930	3.479	1.939	557
97-98	8.818	7.465	847	3.372	1.150	341
98-99	8.865	7.729	872	3.553	2.025	570
99-2K	8.223	7.793	948	3.194	1.753	549
2K-01	8.330	7.027	844	3.695	2.394	648
01-02	8.410	8.170	971	4.797	2.816	587
02-03	8.725	7.791	893	4.798	2.519	525
03-04	7.606	7.871	1035	5.095	1.711	336
04-05	7.386	6.722	910	4.180	2.240	536
05-06	7.396	6.606	893	4.180	2.290	548
06-07	6.020	6.530	1085	6.510	3.800	473
07-08	6.300	9.120	1448	6.05	3.67	607
08-09	6.100	8.600	1412	7.30	4.00	552
09-10	6.500	8.500	1308	9.70	5.70	551

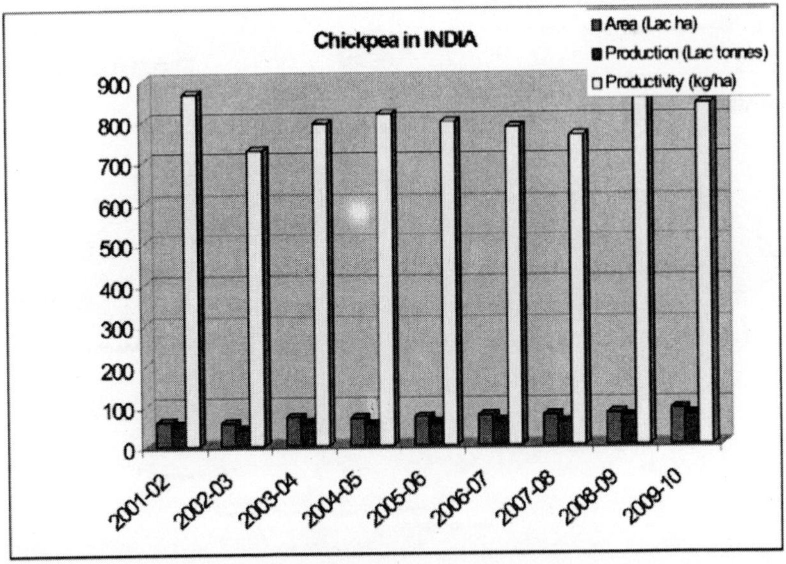

Fig. 1.1. Area (lac ha), production (lac tonnes) and productivity (kg/ha) of Chickpea in India

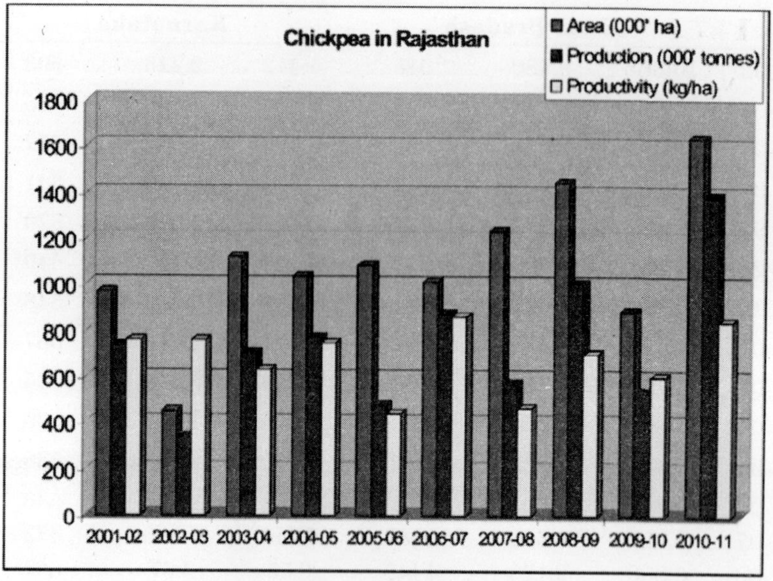

Fig. 1.2. Area (th /ha), production (th/tonnes) and productivity (kg/ha) of Chickpea in Rajasthan

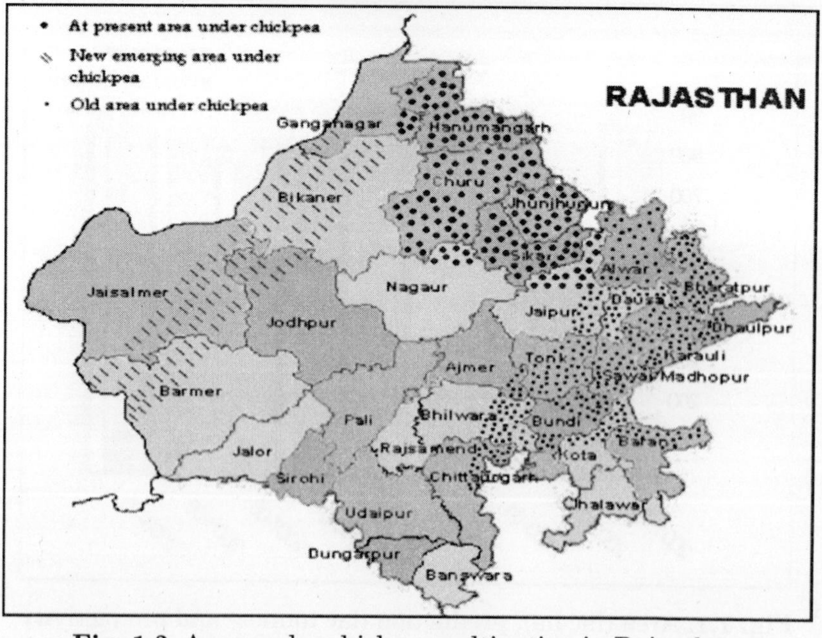

Fig. 1.3. Area under chickpea cultivation in Rajasthan

REFERENCES

Annual Report, 2007-08. All India Coordinated Research Project, ARS Durgapura, Jaipur Rajasthan.

Ali, Masood and Mishra, J.P. 2004. *Rabi* pulses (Chickpea, Lentil, Lathyrus and French bean). Text book of field crops production. Directorate of information and publication of Agriculture by Rajendra Kumar. ICAR Krishi Anusandhan Bhavan, Pusa, New Delhi. **PP:** 317-371.

Janoria, M.P.; Gour, V.K. and Singh, C.B. 1984. Perspectives in grain legumes. National Seminar on "New Dimensions in Pulse Research and Development" held at Jabalpur.

Khedar, O.P. and Maloo, S.R. 2012. Seed production in pulses. Pulses produc tion Strategies. by Dr. P. Murugesa Bhoopati and Dr. M. Paramathma. Sri Shakthi Promotional Litho Process. S.F. No. 283 Masaniamman Nagar, Anna Nagar East, Edarpalayam, Coimbatore- 641025. **PP:** 347-362.

Khedar, O.P.; Singh, R.V.; Shrimali, Mahesh and Singh, N.P. 2008. Pulses: Status and Cultivation Technology. Aaviskar Publishers, Distributors, Jaipur (Raj.)

Mishra, S.P. 2004. Project Coordinator's Report. All India Coordinated Research Project on Chickpea. Indian Institute of Pulses Research, Kanpur. Annual Group meet was held at JAU, Junagarh, Sept. 15-17, 2004.

Sekhon, H.S.; Singh, Guriqbal and Hari Ram 2007. Scope of pulses in crop diversification. Pulses at a glance. Pulses section, Department of Plant Breeding, Genetics and Biotechnology, PAU Ludhiana.

Singh, N.P.; Sewak, Shiv and Iquebal, Mir Asif 2009. All India Coordinated Research Project on Chickpea: A Profile IIPR, Kanpur.

Singh, N.P; 2011-12, All India Coordinated Research Project on Chickpea, IIPR, Kanpur. **PP:** 31-37

Singh, S.J. 2001. Pulses production in Rajasthan. Souvenir, "Annual Group meets of AICRP on Chickpea and MULLaRP" was held at Agriculture Research Station, Durgapura, Jaipur, on Sept 3-6, 2001. **PP:** 10-13.

2

Constraints and Strategies in Chickpea Cultivation

S.S. Shekhawat, O.P. Khedar and S.S. Punia

CONSTRAINTS IN CHICKPEA CULTIVATION

In spite of great importance of chickpea in human diet and Indian economy, the production has not yet increased proportionately. Over the past half century, though, the productivity of chickpea has increased from 651 (1964-65) to 841 (2009-10) kg/ha but even then, we are far behind as per our expectation. Looking to the situation it is important to analyze the constraints in pulses production. The main constraints are:

1. Needs more energy

Chickpea is rich in proteins and needs nearly 2.5 times more energy for per unit production in comparison to cereals, which are rich in carbohydrates or in chickpea ultimate product is protein so lot of energy required in conversation of carbohydrates to protein. That's why it is well known fact that we cannot increase the productivity as we have achieved in cereals.

2. Unfavorable weather conditions

In general chickpea is cultivated in rain-fed condition. So the productivity is largely depends on weather condition. In general there is scarcity of moisture during the crop season results in low yield. The rise in the temperature at the time of maturity leads to forced maturity and decreases the yield levels. Frost and low temperature during the night cause heavy damage to chickpea in Northern India.

3. Neglected soils for cultivation

Generally pulses are cultivated on marginal and sub-marginal soil, which are poor in fertility and water retention capacity, resulting low productivity of pulses. Mostly the pH of such soil and water is more than 9.0 and SAR of water is more than 10.0 which affects the yield badly because pulses are more sensitive to acidity, alkalinity and excess moisture.

4. Varietal constraints

Though, a number of improved varieties are available in chickpea, even then, it is also true that there is lack of high yielding varieties, which are resistant to insect-pests, diseases, cold, drought and are suitable for unfavorable weather conditions.

5. Seed constraints

The seed replacement rate of improved pulses varieties is very low, resulting use of poor quality seed. The inadequate availability of seed of improved varieties continued to be a constraint for increasing chickpea production. Higher seed rate and poor economic condition of chickpea cultivators are also some of the reasons due to which cultivators are unable to use improved seeds of pulses. In sowing schedule Chickpea are given last preference. The availability of improved seed is also not proper like cereals, so seed producers and seed sellers are also least interested in chickpea seed due to uncertain marketing.

6. Neglected in the use of manures, fertilizers and cultures

Primarily they are grown on low fertility land without fertilizer application. Due to poor condition of farmers either they do not use or they use very small quantity of manures, fertilizers and cultures, resulting low yield of pulses. Chickpea crop fix atmospheric nitrogen by themselves through nodules in which indirectly reduces the yield.

a. Manures

It is a fact that most of the cultivators do not use manures in the chickpea crop. It is true that pulses do not require much manure, but if sometimes soil are so poor, application of manure is essential.

b. Nitrogen

Though, chickpea crop is able to fix atmospheric nitrogen to the soil through symbiosis by the help of *Rhizobium*. However, before the plant starts the nitrogen fixation, a starter dose of nitrogen is essential for initial growth of crop, but in general the cultivators do not use nitrogenous fertilizer.

c. Phosphorous

Due to its key role in vital energy transformation of living tissues, it is called as the *mineral of life*. It is an essential nutrient of all living beings. Though, pulses fix the atmospheric nitrogen to the soil but no such source is available for phosphorous. In case of phosphorous PSB only converts insoluble phosphorous in the insoluble form. Hence, in pulse growing soils phosphorous deficiency is increasing day by day.

d. Sulphur

Oil seed and pulses needs more sulphur than cereals. Sulphur is indispensable for the synthesis of amino acids like methionine and cystine. In general the sulphur uptake by pulse crops is 8 kg S/tonne of grain in comparison to cereals (3-4 kg S/tonne of grain). So, sulphur deficiency is also increasing in pulse growing areas day by day.

e. Calcium

For normal growth of the plant, application of calcium is essential. It is recognized more widely as a conditioner of soil than essential plant nutrient. It is constituent of cell wall in the form of calcium pectate, which is necessary for normal cell division.

f. Trace elements

Though, trace elements are required in a very small quantity, but they are also very much essential and their deficiency cause many deformities and diseases. Intensive cultivation causes deficiency of trace elements to the soil. Till date farmers are not aware about the use of trace elements too.

g. *Rhizobium*, PSB and PGPR

Production of efficient *Rhizobium* strains and distribution are not adequate. The use of *Rhizobium*, PSB and PGPR helps in increasing the yield of chickpea. *Rhizobium* fixes the atmospheric nitrogen to the soil through symbiosis, PSB converts insoluble phosphates in soluble form and PGPR promotes the plant growth. Un-awareness or un-availability is the main draw back in the use of cultures. Though, some of the farmers use the cultures, but due to improper storage of culture, strains do not survive and results in low yields.

7. Agronomic constraints

a. Improper sowing

Most of the Chickpea are photosensitive so that growth and yield are affected by time of sowing. Pulse crops get last priority in sowing schedule, results in poor growth and more attack of weeds, diseases and insect-pests. In general broadcasting method is used for sowing, resulting insufficient plant population.

b. Low seed rate

Due to high seed rate and higher prices of pulses seed, farmers generally use low seed rate due to their poor economic condition, results in low productivity. Sometimes farmers use mix cropping in which limited seed of is used.

c. Improper method of fertilizer application

Chickpea respond well to P_2O_5 if the fertilizer placed deep in the feeding zone only at the time of sowing. Farmers either do not apply fertilizer or if apply they do not apply properly. In general they broadcast the fertilizer which is not desirable and plant is not able to take even 25% of the fertilizer used, results in low yields.

d. Inadequate inter culture operations & weed control

Pulses are last priority in case of inter-culture operation and weeds stand in competition to the wheat or barley. Weed management is very much important because chick-pea is short statured so if weeds are not removed on time, they will cover the crop and affects badly to the yield. Soil aeration in such condition is also not proper which results in low yields.

e. Insufficient irrigations

In general chickpea is cultivated in rain-fed areas where irrigation facilities are not available; in such situation crop often face moisture stress at various growth stages. The areas where irrigation facilities are there, farmers irrigate chickpea in case of surplus water only or sometimes they don't irrigate the crop even in case of surplus water.

f. Untimely harvesting

Chickpea crop should be harvested at proper time and if it is delayed there may be heavy losses by shattering of pods in the fields.

8. Plant protection constraints

Now-a-days the biotic stresses are big challenge in chickpea production. Chickpea attacked by so many diseases, nematodes and insect-pests causing heavy loss to the crop. Chickpea suffer more in comparison to cereals from diseases, insect-pests and nematodes. Diseases are serious problem in chickpea and sometimes occur in epidemic form, resulting total failure of the crop. Among the various root-knot nematodes, *Meloidogyne incognita* and *M. javanica* also emerging one of the major constraint reducing both quality and quantity of pulse crops. These are hidden enemy causing heavy losses to the crop. Nematode causes 5-60% yield losses. Insect-pests are again major biotic constraint in chickpea cultivation because chickpea is attacked by so many insect-pests starting from sowing up to consumption. Gram pod borer is major threat in chickpea cultivation. Indian farmers due to their poor economic condition and improper knowledge unable to adopt proper plant protection measures, result in low yields.

9. Post harvest losses

Though, post harvest losses are big problem in all the agricultural products but it is more in pulses in comparison to cereals, because pulses are very much sensitive to bruchid causing an estimated loss of nearly 0.21 million tonnes of pulses, costing 315 million rupees. In chickpea bruchid infection is carried out from fields, whereas, in cereals insects do not come from the fields. Together this, farmers do not have sufficient storage facility, which also results in heavy loss due to storage pests and rats. The post harvest losses are also due to lack of scientific post harvest management system. In such condition farmers are compelled to sell their produce to the middleman at low price which discourage the farmers to grow chickpea.

10. Extension service constraint

The technology gap between research institutions and farmers is very wide. Our extension services are not sufficient and unable to transfer even half of the technology generated by our scientists. Unless the scientific technology is transferred to the farmers through demonstrations, kisan mela, media, audio-visual aids or by any method, it is quite impossible to enhance chickpea production in the country. Poor transfer or adoption of improved technology results in low production of chickpea.

11. Socio economic condition of farmers

Due to poor economic condition and insufficient credit facilities, farmers are unable to purchase improved quality seed, implements, fertilizers and chemicals for plant protection. They are also unable to develop their own irrigation system and are not aware with latest scientific knowledge about cultivation.

12. Fluctuation in market prices and poor post harvest technology

Fluctuating market prices and Poor post harvest technology is again responsible for low production of chickpea. Though, there is high fluctuation in prices of all the agricultural products but chickpea suffers badly. The prices of chickpea fluctuate more in comparison to cereals, which cause uncertainty of income and discourage the cultivation of pulses. Poor post harvest technology is also responsible for low production of chickpea. In the absence of proper post harvest technology sometimes the production get spoiled.

13. Risky and less remunerative than cereals in intensive cultivation

Cereals are more remunerative and less risky than chickpea under intensive management practices, results in the decline of net sown area.

14. Green revolution

Food grain shortages of the mid-sixties, green revolution and changes in the infrastructures and incentives including input supplies and price support systems in favour of major cereals altered the traditional cropping pattern against pulses. With decades of emphasis on research as well as production of superior cereals, chickpea/pulses were driven out of not only the irrigated areas, but also the rainfed farming and were relegated to the marginal lands.

15. Domestic need

Cereals are the staple diet for every human being. Cereals supplying feed and fodder both, so first of all every farmer has to cultivate cereal for the domestic consumption and if there is surplus land then only he will go for chickpea.

STRATEGIES FOR INCREASING CHICKPEA PRODUCTION

As pulses play an important role in Indian economy, human diet, cattle feed, agricultural export and nutritional security of poor, the production of the pulses in the country should be enhanced. For increasing pulse production we have to pay attention on following points:

1. Cultivation on better soils

Motivate the farmers for cultivation on fairly better lands and timely sowing. Introduction of chickpea in irrigated farming system is again important for improving production and productivity of chickpea. Now-a-days chickpea is not poor man's food only, but they are added in the dietary habit of rich peoples also. Chickpea is more remunerative than cereals due to their high nutritive value and demand in global and domestic markets.

2. Development of improved varieties

Though, our scientists have developed so many varieties which are of short duration, high yielder and resistant up to some extent to the diseases and insect-pests, but it does not full-fill our requirement. It is a fact that a variety which is resistant to a particular disease or insect-pest today may be susceptible in the future. Now-a-days we have to develop varieties for a problem and region specific such as the areas where there is attack of diseases and insect-pests, we have to develop resistant varieties for that particular disease or insect-pest. The areas where rain-fall is very low we have to develop drought resistant and short duration varieties. In the same way if we want to export, we have to develop quality product with high protein. Now-a-days due to high price of produce, farmers are growing chickpea under better management condition too, so genotypes with high management should be developed.

3. Proper supply and use of improved variety seed for sowing

Though, majority of the farmers are very cautious about the use of improved seed of cereals but they are less aware about improved seed of pulses till date. Till today the seed replacement rate in chickpea is very less, which is a serious issue and by improving the seed replacement rate we can increase the pulse production. Again, it is very essential to provide the seed at reasonable price to the farmers because the seed rate of pulses is very high and in most of the cases price of seed is a major constraint in pulse production.

4. Use of proper fertilizers, manure & trace elements

Though, chickpea do not show much response to nitrogen only to starter dose due to nitrogen fixation ability, whereas, they have shown significant response to phosphorus. The recommended dose of fertilizer (NPK) helps in enhancing production. It has been observed that most of the pulse crops respond well to 20 kg sulphur/ha and in some areas with light textured soil 40 kg/ha also respond well. Gypsum, pyrite or SSP are the low cost source of sulphur. Application of zinc, iron, molybdenum and boron also respond well in pulses production. Foliar spray of 2% urea at the time of flowering/pod filling stage results in higher yield. With the introduction of high yielding varieties, it is very essential to apply balanced fertilizers to pulse crops.

5. Proper seed treatment

It is a fact that a little investment on seed treatment enhances chickpea production up to great extent. Proper seed treatment with fungicide and insecticide helps to overcome diseases and insect-pests so seed should be treated properly by the recommended dose of chemicals.

6. Use of *Rhizobium* culture, PSB and PGPR

Seed inoculation with efficient strains of *Rhizobium* culture PSB and PGPR is the cheapest input in pulse cultivation, which increases the yield significantly. *Rhizobium* fixes the atmospheric nitrogen to the soil, PSB converts the insoluble phosphates in to soluble form and make them available to the plant and PGPR enhances the growth of the plant. So, *Rhizobium* culture, PSB and PGPR each @ 600 g/ha should be used for seed treatment in pulses for better production.

7. Use of trace elements

Though, the trace elements are required in very little quantity but they increase the yields significantly, because the absence of trace elements in the plants affects so many physiological activities badly and results in low yields or sometimes attacked by diseases or insect-pests resulting total failure of the crop. For good production of chickpea required quantity of trace element should be applied. Use of micronutrients in problematic soil also helps in good production.

8. Improvement in agronomical practices

For better production agronomical practices from field preparation to harvesting should be done on proper time.

Sowing time, nutrient management, weed management, irrigation and timely harvesting are some of the important agronomic practices, which affect the production up to a great extent. Crop rotation is also an important tool for enhancing productivity. Use of optimum seed rate and maintain optimum plant population. Cotton-chickpea, rice-chickpea, pearlmillet-chickpea, sorghum-chickpea, maize-chickpea are some of the crop rotations which will be helpful in increasing production of chickpea. In absence of rains, application of irrigation helps in good yield. Effort should be made to divert and collect the rainwater at appropriate site for future use as life saving and protective irrigation. Keeping the field free from weeds up to 60 days helps in improving chickpea production. Installation of sprinkler set to bring more area under irrigation will also help in increasing the area under chickpea.

9. Adoption of proper plant protection measures

Chickpea crop is sensitive to diseases, nematodes and insect-pests. Diseases or insect-pests sometimes occur in epidemic form and cause total failure of the crop. Absence or delay in plant protection may cause great losses in yield. Application of appropriate plant protection module at right time results in good crops. For management of biotic stress integrated modules for crop protection should be developed well in time.

10. Improvement in extension services

Extension agencies may play a major role in agricultural production and its quality both. It is well known fact that till date the technology developed by the scientists has not reached to the cultivators. There is wide gap between the technology generated and adopted by the farmers. To bridge the gap it is essential that it may be transferred through extension agencies. Demonstrating improved technology through demonstrations or mini kits is an effective way. In case of chickpea the seed replacement rate of improved varieties is very low which should be improved. In the same way sometimes some of the diseases and insect-pests occur in epidemic form resulting total failure of crops so, our extension services should be improved and latest technical guidance with inputs should be made available to the farmers at their fields. Time to time training, field demonstration and farmer tours should also be organized for farmers.

11. Improvement in socio-economic status of the farmers

It is very sad that a country like India, which is known by its villages and economy is also based on agriculture, the socio-economic status of its farmers is very poor. For a developing country of 21st century it is very essential to improve the socio-economic status of its back bone the 'Farmer' by providing the subsidies on inputs (like seed, fertilizer, chemicals, implements, electricity charges, irrigation system, irrigation charges of canal water etc.), easy finance, crop insurance, road network up to mandies, hygienic living condition and proper education to their children. If the economic status of farmers will be improved, it is guaranteed that farmers will be able to use improved varieties, essential inputs, plant protection measures for cultivation of chickpea. This will definitely improve the chickpea production of the country.

12. Improvement in post harvest technology and value addition

Pulses are more sensitive to store grain pests, rats and even cattles in comparison to cereals. According to various estimates, total post harvest losses are to the tune of 20-25% out of which milling alone contribute 10%. We have to overcome these losses and save the precious food for human and animals. To overcome post harvest losses our government or other organization should come forward to provide storage facilities to the farmers on nominal rates so that the farmers may put their produce safely up to a desired time.

To overcome insect infestation management strategy can be planned viz. eliminate source of infestation in field and store, reduce grain moisture through sun drying, use of indigenous storage structures, improvement of mud bins and plastic lined bags, use of mechanical traps and sex pheromones, use of perspective material like oils, plant products, inert material and synthetic insecticides, use of proper storage and grain pest resistant varieties. A mass awareness programme to educate farmers on scientific storage can also check the post harvest losses up to great extent. There is an urgent need to develop post harvest technologies to promote on-farm processing and value addition through development of energy efficient pre-treatment processes. To improve value addition and agro-processing activities, distribution of *dal* mills and information on final products and by-products of pulses definitely improve the pulse production. This would also protect the resource-poor farmers from the hands of middlemen and market forces as pulse farmer do not get even half of what consumer pays. Several processing techniques like chemical detoxification, fermentation, soaking, cooking, dehulling, milling, roasting and boiling needs to be standardized to make pulses more palatable and less toxic so that more demand of pulses will be created and it will definitely enhance the pulse production in the country.

13. Improvement in availability of farm machines

Farm machinery or implements are very essential for successful cultivation of a crop. For agricultural operations proper implement or machinery should be made available to the farmers at reasonable price, so they can use machines and may improve their working efficiency. This will definitely encourage the chickpea cultivators.

14. Policy Support

The strong infrastructural support for input supply, credit and assured marketing in the rain-fed areas will be helpful in increasing chickpea production. Input should be supplied well on time at reasonable price and required place. Easy credit facility should be given according to their need at reasonable interest, support price system should be made more practicable and assured marketing facility should be given to the pulse cultivators. Minimum support price (MSP) should be declared before the sowing time so that farmers can plan their sowing well in time. Furthermore, farmers should be encouraged by reasonably increasing the minimum support prices of the pulses to produce more pulses. Crop insurance is also helpful in improving pulse production because cultivation of pulses is more risky in comparison to cereals.

15. Additional area under pulses

With the development of short duration varieties, resistant to insect-pest and diseases and drought resistant varieties there is tremendous scope for expansion of area. Short duration varieties of pulses can fit well in various cropping systems. In addition, there is a scope for introduction of pulses in niches such as wasteland, reclaimed soils and rice-fellow by efficient watershed management and as a replacement of less remunerative crops. The scarcity of irrigation water is again major cause of increase in area. Due to lack of sufficient irrigation farmers are shifting to chickpea cultivation.

REFERENCES

Ali, Masood and Mishra, J.P. 2004. *Rabi* pulses (Chickpea, lentil, lathyrus and frenchbean). Text book of field crops production. Directorate of information and publication of Agriculture by Rajendra Kumar. ICAR Krishi Anusandhan Bhawan, Pusa, New Delhi. **PP:** 317-371.

Bahl, P.N. 1988. Breeding approaches for self pollinated crops "Grain Legumes". Plant Breeding by V.L. Chopra. Oxford and IBH Publishing Co. Pvt. Ltd. New Delhi, Mumbai and Kolkata. **PP:** 121-137.

Janoria, M.P.; Gour, V.K. and Singh, C.B. 1984. Perspectives in grain legumes. National Seminar on "New Dimensions in Pulse Research and Development"was held at Jabalpur.

Khedar, O.P. and Maloo, S.R. 2012. Seed Production in Pulses. Pulses Production Strategies. by P. Murugesa Bhoopati and M. Paramathma. Sri Shakthi Promotional Litho Process. S.F. No. 283 Masaniamman Nagar, Anna Nagar East, Edarpalayam, Coimbatore- 641025. **PP:** 347-362.

Khedar, O.P.; Maloo, S.R. and Punia, S.S. 2012. Pulses in Rajasthan: An overview. Pulses production Strategies. by P. Murugesa Bhoopati and M. Paramathma. Sri Shakthi Promotional Litho Process. S.F. No. 283 Masaniamman Nagar, Anna Nagar East, Edarpalayam, Coimbatore- 641025. **PP:** 191-226.

Khedar, O.P.; Singh, R.V.; Shrimali, Mahesh and Singh, N.P. 2008. Pulses: Status and Cultivation Technology. Aaviskar Publishers & Distributors, Jaipur (Raj.)

Mishra, S.P. 2004. Project Coordinator's Report. All India Coordinated Research Project on Chickpea. Indian Institute of Pulses Research, Kanpur. Annual Group meet was held at JAU, Junagarh, Sept. 15-17, 2004.

Singh, N.P. and Sewak, Shiv 2007. Recent achievements in chickpea research. National Symposium on Legumes for Ecological Sustainability: Emerging Challenges and Opportunities held at IIPR, Kanpur, Nov. 3-5, 2007 **PP:** 49-55.

Singh, N.P.; Sewak, Shiv and Iquebal, Mir Asif 2009. All India Coordinated Research Project on Chickpea: A Profile IIPR, Kanpur.

3

Chickpea Breeding

S.S. Shekhawat

BOTANY OF PLANT

Chickpea (*Cicer arietinum* L.) belongs to the family leguminosae and sub-family papilionaceae with the chromosome number 2n = 16. It is a freely branched annual herb, 25 to 50 cm tall with erect or somewhat spreading growth. Strong having tap root system and lateral roots with large lobed nodules. Leaves yellow to bluish green in colour, stipules ovate 8 mm long; leaflets ovate, elliptic or obviate, serrate, up to 2 x 1.5 cm. Inflorescence solitary. Flower papilionaceous, zygomorphic and hermaphrodite; sepals lanceolate, gamosepalous; petals greenish white to pink or blue, standard broad and clawed, wing free, keel incurved; stamens diadelphous (9 + 1); ovary superior, style filiform, beardless with terminal stigma. Pod inflates up to 3 x 2 cm, 1 or 2 seeded and occasionally 3 seeded. Seeds angular, prominent beak and small hylum, colour white brown, light red to black. Germination hypogeal. Cotyledons are thick and yellowish.

BREEDING OBJECTIVES

1. Development of high yielding varieties suitable for rain fed conditions, resistant to drought, dry root rot, pod borer and nematodes.
2. Development of high yielding, short duration varieties suitable for different agro-climatic conditions.
3. Breeding for better grain quality, value addition and consumer preference.
4. Breeding for extra large seeded 'Kabuli' varieties suitable for export purpose.
5. Breeding for high input responsive varieties.

IMPORTANCE OF QUALITY SEED

Seed is fertilized and ripened ovule contains a viable embryo, able to germinate and produce same off-spring. Seed is carrier of new technologies or a basic tool for secured food supply. Seed is the most important input of agriculture and its importance is well known in agriculture because:

1. It is the backbone of sustainable agriculture and is initial and the basic input responsible for improving productivity and quality of produce.
2. Seed remains a key input in crop production.
3. It is the basic unit of cultivation.
4. It carries the genetic potential of the crop.
5. It is a source of continuity.
6. It plays a vital role in accelerating and sustaining crop productivity.
7. Seed dictates the ultimate productivity of other inputs such as fertilizers, irrigation, pesticides etc. are able to give higher yields only when good seeds are used.
8. It is the cheapest and most effective agricultural input.

Till date it is true that the availability of quality seed in chickpea is a major constraint and it is also true that without proper supply of quality seed it is quite impossible to get desired production. Unfortunately, the productivity is very low, mainly due to poor quality seed. Plant breeders generally take 10-12 years to develop (selection, testing and release) a variety by traditional plant breeding. The worth of a variety can be realized by farmers only if sufficient quantity of quality seed is made available to them. The production and supply of quality seed is far less than the required quantity to achieve the desired production levels. The seeds replacement rate is very less (5-8 per cent). Indian economy is mainly depends on agriculture which is mainly dependent on quality seeds. *Seed being a biological and vital input plays a pivotal role in the development of agriculture in any country.* It acts as a catalyst for realizing the potentiality of other inputs. *To increase agricultural production, viability of quality seed is one of the pre requisite.* Neither the poor quality seed of superior variety nor the high quality seed of inferior variety can serve the farmer satisfactorily. Thus, the basic objective of any seed production programme is production and distribution of quality seed. Under the traditional system of farming most of the farmers used to retain a part of their produce for use as seed. Advances in research resulting in the evolution of high yielding, short duration, dwarf, drought tolerant and disease and insect-pest resistant varieties have added a new dimension to the chickpea production.

Though, there are so many reasons of low productivity of chickpea like rain-fed cultivation, sowing on marginal lands, improper agronomic practices, so many abiotic and biotic stresses and inferior storage facilities but, use of poor quality seed is most important factor responsible for low production and productivity of legumes. The quality seed is the cheapest input with maximum return. It is estimated that quality seed of improved varieties can contribute about 15-20% increase in production. Therefore, it is true that there is no substitute of this input and emphasis should be given on the quality seed production of legumes. The need of quality seed having desired characteristics is well recognized for increasing the crop yield in any agro-climatic region. Though, Indian seed industry has made outstanding contribution in production and timely supply of quality seeds to the farmers even than the seed supply is hardly 15-20 per cent and if we talk about pulses it is only 5-10 per cent which is a serious issue. The rest of the seed come from farmer's saved seed which is inferior one and responsible for low production. Quality seed production is a skilled job and requires several standards to be meet-out judiciously. *The quality seed development starts from selection of land, agronomy, plant protection, harvesting, threshing, seed processing and storage. The quality seed production* differs from commercial crop production in several respects *and* special methods and precautions are needed to produce seed of desired quality. It needs *care and vigilance at each and every stage.* Isolation distance, monitoring of seed and post harvesting handling of seed are some of the factors affecting seed quality so in any seed production programme we have to take care about these points.

The availability of quality seed of improved varieties to the farmers at an affordable price and in time is considered crucial for enhancing and sustaining the agricultural productivity and economic growth. Only assured quality can respond to other inputs otherwise seed of hope may turn into seed of frustration. The role of seed is well known in agriculture. At farm level, quality seed results in enhanced productivity, reduced risk, increased net income through higher yields, more efficient use of available nutrients, faster maturation, better resistance to insect-pests, higher nutrients content in harvested crop, whereas, at national level more diversified agricultural production system are made possible by multiple cropping and spread of crops to wider set of agro ecological zones.

VARIETIES IN SEED CHAIN

The varieties in seed chain are A1 (Annegeri 1), Avrodhi, B 108 (Mahamaya I), B 115 (Mahamaya II), BG 372 (Pusa 372), BG 391, BG 1003, BG 1053, BGD 72, BGD 128, BGM 547, C 235, CSJD 884 (Akash), Dahod Yellow, DCP 92-3, Digvijay, GCP 101 (Guj. Gram 1), GNG 469 (Samrat), GNG 663 (Vardan), GNG 1292, GNG 1488 (Sangam), GNG 1499 (Gauri), GNG 1581 (Ganguar), GPF 2, Gujarat Gram 2, Gujarat Gram 4, HC 3, HC 5, Himachal Chana 2, HK Chana 2, H. Kabuli Chana 1, ICCV 2 (Sweta), ICCV 10 (Bharti), ICCC 37 (Kranti), JAKI 9218, JG 11, JG 16 (SAKI 9616), JG 63, JG 74, JG 130 (Jawahar Gram 130), JG 218, JG 315, JGG 1, JKG 1, KAK 2 (PKV 2), KGD 1168 (Alok), KPG 59, Pant G 114, L 550, Pant G 186, PBG 1, PBG 5, PDG 4, Phule G 5, Phule G 12, Phule G 9425-9 (Rajas), Pragati, Pratap Chana 1, Pusa 256, Pusa 1088, Pusa 1103, Pusa 1105, Pusa 1108, Radhey, RSG 44, RSG 807 (Abhar), RSG 888, RSG 896 (Arpan), RSG 902 (Aruna), RSG 931, RSG 945 (Asha), RSG 963, RSG 973 Abha), RSG 991 (Aparna), RSG 995 (Arpita), Vaibhav, Vijay, Virat, Vishal, WBG 39/2 (Anuradha), WCG 1, WCG 2 (Surya), etc.

IMPORTANT RELEASED VARIETIES

The important varieties, their pedigree, year of release, days to maturity, average yield, area of adaptation, reaction to measure diseases and special characteristics are given in Table 3.1.

Table 3.1. Chickpea varieties their pedigree and characteristics

S.N	Name of variety	Developed from	Pedigree	Year of release	Avg. yield (q/ha)	Days to maturity	Area of adaptation	Salient characteristics
1.	C 235 (G 235)	PAU, Ludhiana	IP 58 x C 1234	1972	20	-	Haryana & Punjab	SRA, SS (13 g/ 100-seed), tol. to *Ascochyta* blight
2.	ICCC 32 (Kabuli)	ICRISAT, Patancheru	L 550 x L 2	1984	26	-	NWPZ	WR, tol. to root rot & pod borer

3.	Pusa 244 (BG 244)	IARI, New Delhi	(850-3/27 x P 922) x P 9847	1985	18-22	110-120	CZ	Tall, leaf ovate with serrate margins, seed light brown & bold, ABR, res. to stunt, tol. to cold
4.	Pusa 256 (BG 256)	IARI, New Delhi	[(JG 62 x 850-3/27)] x [(L 550 x H 208)]	1985	18-20	-	All India	Medium tall, leaf broad, seeds light brown and attractive bold, tol. to *Ascochyta* blight and wilt
5.	Pusa 261 (BG 261)	IARI, New Delhi	P 827 x P 9847	1985	20-25	-	NWPZ	Tall, Erect & compact, seeds golden yellow and medium bold, tol. to *Ascochyta* blight
6.	Pusa 417 (Girnar)/ (BGM 417)	IARI, New Delhi	Induced mutant of BG 203	1985	20	140-145	CZ	Suitable to normal and late sown condition, WR, RRR, CRR, SS with golden yellow colour
7.	Vishwas (Phule G 5)	MPKV, Rahuri	B 110 x N 31	1986	20-25	-	CZ	Tall and erect, seeds smooth, bold & round, tol to wilt and pod borer
8.	Pusa 267 (BG 267) (Kabuli)	IARI, New Delhi	US 613 x BEG 482	1987	22	140-150	NWPZ	Tall & semi erect, seeds solmon white and medium bold
9.	SG 2	Dholi, Bihar	(E 100 x P 436) x (L 550 x F 378)	1987	20-22	140-145	South Bihar	Double podded and tol. to wilt
10.	PBG 1	PAU, Ludhiana	GG 578 x NEC 206	1988	16	155-160	NWPZ	Plants tall and erect, seeds brown and small, tol. to *Ascochyta* blight
11.	Gora Hisari (Kabuli)	HAU, Hisar	-	1988	15-16	-	Haryana	Semi erect, SS with white colour
12.	Kranti (ICCC 37)	ICRISAT, Patancheru	(P 481 x JG 62) x P 1630	1989	16-20	98	CZ & SZ	Dwarf, leaves light green and narrow, Semi-erect, seeds light brown

13.	Harayana Chana 1 (H 82-2)	HAU, Hisar	F 61 x L 550	1990	22	145-150	NWPZ	Dwarf, FWR, SS, Suitable for normal and late sown condition
14.	Phule G 12	MPKV, Rahuri	GW5/7 x Ceylon	1990	21	96-100	MS	Res. to wilt and tol. to pod borer
15.	RSG 44	ARS, Durgapura	JG 62 x F 496	1991	15-18	135-150	Raj.	Double podded, tol. to drought & frost, sui. for rainfed, irrigated and late sowning
16.	Bharati (ICCV 10)	ICRISAT, Patancheru	P 1231 x P 1265	1992	20	96	CZ and SZ	Semi-erect, FWR, Dry RRR
17.	Sadabahar	CSUA&T, Kanpur	Hima x L 245	1992	25	140-150	UP	Green seeded and tol. to wilt
18.	Udai (KPG 59)	Bharari	-	1992	20	120-140	NEPZ, NWPZ	Semi-erect, medium tall, small seeded, tol. to root rot, wilt, stunt and pod borer
19.	Pusa 329 (BG 329)	IARI, New Delhi	BG 203 x P 179	1993	22	145-155	NWPZ	Errect, seeds brown and bold, mod. Res. to wilt
20.	Pusa 372 (BG 372)	IARI, New Delhi	P 1231 x P 1265	1993	14-21	-	NWPZ, NEPZ & CZ	Semi spreading, mod. Res. to wilt, blight & root rot, suitable for late sown condition
21.	Sweta (ICCV 2) (Kabuli)	ICRISAT, Patancheru	[(K 850 x GW 5/6) x P48] x (L 550 x Guamchil 2)	1993	13	80-90	MS & AP	Semi-spreading, seeds white and bold, res. to wilt and *botrytis* grey molud
22.	Vardan (GNG 663)	ARS, Sri Ganganagar (Raj.)	GNG 16 x GNG 146	1994	24	150	NWPZ	Semi-erect, medium dark brown seeds, WR
23.	Vijay (Phule G 81-1-1)	MPKV, Rahuri	P 127 x Annegeri-1	1994	20	105-110	CZ	Suitable for rain-fed condition, tol. to terminal moisture stress
24.	GPF 2 (GF 89-36)	ARS, Faridkot, Punjab	GL 769 x H 75-35	1995	21	-	NWPZ	Medium tall, semi erect, tol. to *Ascochyta* blight

25.	Pusa 362 (BG 362)	IARI, New Delhi	BG 303 x P 179	1995	24	150	NWPZ	Medium tall, seeds bold, tol. to wilt
26.	GNG 469	ARS, Sri Ganga-nagar (Raj.)	Annegeri x H 75-35	1996	21	147	NWPZ	Tall and erect, seeds brown and bold, tol. to wilt and root rot, suitable for irrigated and rainfed areas
27.	Pusa 391 (BG 391)	IARI, New Delhi	ICC 3935 x Pusa 256	1996	17-18	110-120	CZ	Medium tall, erect, seeds bold and dark brown, mod. res to wilt and root rot
28.	Hirwa Chaffa (AKGS 1)	PDKV, Akola (MS)	Sel. From local GP	1996	16	105-110	MS	Green seeded, WR
29.	Pant G 186	GBPUA&T Pantnagar	ILG 613 x Pant G 114	1996	18	-	UP	Small seeded and suitable for late sown condition
30.	JG 218	JNKVV, Jabalpur	ICCC 4 x P 1353	1996	18	115	MP	Early maturing, tol. to wilt
31.	Vishal (Phule G 87207)	MPKV, Rahuri	K 850 x ICCL 80074	1997	20	112	MS	WR and suitable for rainfed areas
32.	Alok (KGD 1168)	Kanpur	Sel. K 158	1997	19-21	140-150	NWPZ	MBS, WR, RRR
33.	JG 322	JNKVV, Jabalpur	Chaffa x BG 1	1997	15-18	100-115	MP	FWR, 100-seed weight 16 g
34.	PDG 3 (GF 89-133, GPF 133)	PAU, Ludhiana	GL 769 x GNG 146	1997	16	160	Punjab	Tol. to pod borer
35.	DCP 92-3	IIPR, Kanpur	Sel. From GP	1998	18-19	150	NWPZ	LR, WR, MBS with yellowish brown colour
36.	Karnal Chana 1 (CSG 8962)	Karnal, Haryana	Sel. From GPF 7035	1998	20-22	150	NWPZ	Suitable for normal as well as saline soils, WR
37.	Chamatkar (BG1053)	IARI, New Delhi	ICCV 3 x FLIP 88-20	1999	17	-	NWPZ	Tall and semi-erect, WBS, WT
38.	Co 3	TNAU, Coimbatore	Sel. Maharashtra Collection	1999	10-11	85	TN	Flowers light pink and veined, BS (30-32 g/100 seed)

39.	Co 4	TNAU, Coimbatore	ICCC 42 x ICCC 12237	1999	11-12	82-85	TN	Flowers light pink and veined, BS (30-32 g/100 seed), seed brown
40.	JGG 1	JNKVV, Jabalpur	Sel. From land races of *gulabi* gram	1999	13-15	130	MP	Suitable for table purpose and has good parching quality, FWR
41.	PKV *Kabuli 2* (KAK 2)	MPKV, Rahuri	-	1999	17-18	91-113	Central India & MS	Bold seeded, semi spreading, flower white, res. to *fusarium* wilt
42.	Pusa *Kabuli* 1003 (BG 1003)	IARI, New Delhi	Mutant of L 532	1999	18	130-140	NEPZ	Medium tall, white bold seeded tol. to wilt
43.	Dharwad Pragati (BGD 72)	IARI, New Delhi	(Pusa 256 x E100 Ym) x Pusa 256	1999	18	115-120	CZ	-
44.	Gujarat Gram 1 (GCP 101)	Junagadh	GCP 2 x ICCV 2	1999	18	115-120	CZ	WR, MBS with dark brown colour seed
45.	Gujarat Gram 2 (GCP 107)	Junagadh	JG 1258 x BDN 9-3	1999	23	95-100	Gujarat	WR, BS
46.	Surya (WCG 2)	GBPUA&T, Modipuram	Mutant of G 130	1999	18-20	125-135	UP	Semi-spreading, foliage light green, flower white, seed small yellowish-brown, mod. Res. to wilt & root rot
47.	JG 11	JNKVV, Jabalpur	(Phule G 5 x Narsingpur bold) x (ICCC 37 ICC x 860263-PB-9P-BP)	1999	-	-	SZ	-
48.	Jaki 92-18	ICRSAT Hyderabad	(ICCC 37 x GW 5/7) x ICCV 107	1997	18	-	CZ	-

49.	Gujarat Gram 4 (GCP 105)	Junagadh, Gujarat	ICCL 84224 x Annegeri	2000	19-20	120-130	NEPZ	Suitable for irrigated and rain-fed areas, WT, Smooth round and brown seeds
50.	SAKI 9516	PDKV Akola & ICRISAT	ICCC 42 x ICCV 10	2000	19-20	-	CZ	Semi spreading, dark green foliage, seeds attractive, light brown and angular in shape, FWR, CRR, SR
51.	Gulak 1 (Gulabi Akola 1)	PDKV Akola, Maharashtra	-	2000	-	-	MS	-
52.	HK 98-134	CCSHAU, Hisar	(H82-5 x E 100 YM) x Bhim	2001	-	-	NEPZ	Kabuli type
53.	RSG 888 (Anubhav)	RAU, ARS Durgapura (Raj.)	RSG 44 x EY00Y	2002	15-20	130-135	NWPZ	Res. to dry root rot & wilt. Seed medium bold & red to dark-brown in colour
54.	Haryana kabuli 1 (HK 89-131	CCSHAU Hisar, Haryana	(L 550 x E 100 Ym) x (ICCC 32 x ICCL 820001)	2002	20	155	Haryana	-
55.	JG 130	JNKVV, Sehore (MP)	(Phule G 5 x Narsinghpur bold) x JG 74	2002	18-20	105-115	MP	BS with angular and smooth surface, FWR, tol. to pod borer
56.	JKG 1	JNKVV, Jabalpur	ICCV 2 x Surutato x ICC 7344	2002	-	-	CZ	Kabuli type
57.	Phule G 95311	MPKV, Rahuri, Maharashtra	ICCC 32 x ICCL 80004) x ICCC 49 x FLIP 82-K) x ICCV 3	2002	-	-	SZ	-
58.	GNG 1292	ARS Sriganganagar, Raj.	Sel. from GNG 149 x (L 550 x L 2)	2002	24	147	Raj.	Kabuli type

	PBG 5	PAU, Ludhiana	BG 257 x E 100 Y	2002	19	165	Punjab	-
59.	HK 98-155 (Haryana channa 3)	HAU, Hisar, Haryana	ICCV 2 x Surotato 77	2003	18-19	145	NWPZ	Spreading type kabuli, Medium tall, white flower, dark green broad leaves
60.	RSG 931 (Anvita)	RAU, ARS Durgapura (Raj.)	RSG 44 x RSG 524	2004	16-17	130-140	NWPZ	Semi erect with profused branching, mod. res. to dry root rot
61.	CSJD 884 (Akash)	RAU, ARS Durgapura (Raj.)	RSG 44 x E 100 Y	2004	15-25	130-135	Raj.	Mod.res. to dry root rot and wilt, seed medium bold and brown
62.	RSGK 6 (Ashar)	RAU, ARS Durgapura (Raj.)	RSGK 628 x ICCV-2	2004	15-20	135-140	Raj.	Kabuli type, mod. res. to dry root rot, res. to nematode
63.	WBG 39/2 (Anuradha)	PORS, Berhampore (WB)	Mahamaya 1 x Radhey	2004	24	126	WB	-
	Pusa 1103	IARI New Delhi	(Pusa 256 x *Cicer reticulatum*) x Pusa 362	2004	22	130	Delhi	Mod. res. against wilt, dry rootrot and stunt, bold seeded 24g/100 seed
64.	HC 5 (H 96-99)	CCSHAU, Hisar	H 89-79 x H 89-84	2004	24	155	Haryana	-
65.	JG 412	JNKVV, Jabalpur	(Phule G 5 x Narsinghpur bold) x ICCV 37	2004	19	95	MP	-
66.	JG 63	JNKVV, Jabalpur	SPS from JG 62	2004	23	115	MP	-
67.	RSG-973 (Abha)	RAU, ARS, Durgapura	RSG- 515 x K-850	2004	15-20	125	Raj.	Res. to dry root rot & nematodes
68.	Digvijay	MPKV, Rahuri	Phule G 91028 x Bheema	2005	17			
69.	RSG 807 (Abhar)	RAU, ARS, Durgapura (Raj.)	BG 256 x RSG 44	2005	20-25	135-140	Raj.	Res. to dry root rot & nematode seed, red-brown
70.	RSG 963 (Adhar)	RAU, ARS Durgapura (Raj.)	RSG 524 x PDG 84-10	2005	20-25	125-130	NWPZ	Res. to nematode, BGM, & stunt

71.	RSG 895 (Arpita)	RAU, ARS Durgapura (Raj.)	RSG 44 x RSG 256	2005	15-20	130-135	Raj.	Res. to dry root rot and wilt
72.	RSG 945 (Asha)	RAU, ARS Durgapura (Raj.)	RSG 668 x RSG 617	2005	20-25	125-130	Raj.	Res. to wilt and nematode
73.	HK 2 (HK 94-134) [Kabuli]	CCSHAU, Hisar	(H82-5 x E 100Ym) x Bheema	2005	22	130	NEPZ	Tol. to wilt and root-rot
74.	Pratap Channa 1	ARS, Kota (Raj.)	ICC 1030 x ICCC 1	2005	13	90	Raj.	Suitable for conserved moisture condition
75.	Pusa 1088	IARI, New Delhi	(BG 256 x ICCV 32) x ICCV 32	2005	23	140	Delhi	Res. to soil borne diseases, stunt, & BGM, tol. to moisture stress
76.	Pusa 1105	IARI, New Delhi	(C104 x BG 1003) x (ICC88503 x BG 1048)	2005	19	145	Delhi/ NCR	-
77.	L 551 (Kabuli)	PAU, Ludhiana	-	2006	18		Punjab	
78.	JG 1	JNKVV, Jabalpur	Sel. from genetic stock	2006	15		MP	
79.	Pusa 547 (BGM 547)	IARI, New Delhi	Mutant of BG 256	2006	18	145	NWPZ	Semi erect plant, tol. to wilt, root-rot and stunt
80.	Phule G 9425-9 (Rajas)	MPKV, Rahuri	Phule G 91028 x Bheema	2006	19	145	NWPZ	Semi erect plant with profuse branching
81.	Pusa Shubhra (BGD 128)	IARI, Dharwad	ICCV 2 x ICCV 5	2006	18	120	CZ	Semi erect, light green foliage, 100-seed weight 27.7g
82.	Pusa 1108	IARI, New Delhi	[(BG 315 x ILC 72) x (ICCV 13 x Flip 85-11)] x (ICCV 32 x Surutoto)	2006	29	150	Delhi	Semi erect with broad and light green leaves, mod. res. against wilt and dry root-rot
83.	JGK 3 (JSC 19)	JNKVV, Jabalpur	ICCV 2 x BG 256	2006	15	115	MP	-

84.	JG 226	JNKVV, Jabalpur	JG 74 x JG 315	2006	21	115	MP	-
85.	JGK 2	JNKVV, Jabalpur	(ICC 32 x L 144) x (ICCC 49 x FLIP 82-16 C) x ICCV x3	2006	18	115	MP	-
86.	RSG 991 (Aparna)	RAU, ARS Durgapura	RSG 289 x BG 1053	2007	16	135	Raj.	-
87.	RSG 896 (Arpan)	ARS, Durgapura	RSG 44 x RSG 538	2007	14	135	Raj.	Res. to wilt and dry root-rot
88.	RSG 902 Aruna)	ARS, Durgapura	RSG 44 x PDG 84-10	2007	20	135	Raj.	Res. to wilt and dry root-rot
89.	GNG 1581 (Gang-uar)	ARS, Srigan-ganagar	GPF 2 x H 82-2	2007	24	151	NWPZ	Medium height semi-erect plant, tol. against wilt, *Ascochyta* blight, stunt & root-rot
90.	GNG 1488 (Sang-am)	ARS, Srigan-ganagar	KPG 279-3 x GNG 469	2007	18	134	-	Medium height semi-erect plant, tol. against pod borer
91.	GNG 1499 (Gauri) [Kabuli]	ARS, Srigan-ganagar	GNG 421 x GB 267	2007	24	143	Raj.	Medium plant height, semi erect, bold seeded
92.	Sadhbha wana	SVBPUAT, Meerut	Mutant of C 235	2008	20	-	UP	-
93.	WCG 2 (Surya)	SVBPUAT, Meerut	Mutant of G 130	2008	21	-	UP	-
94.	WCG 10	SVBPUAT, Meerut	Mutant of G 130	2008	21	-	UP	-
95.	WCG 3 (Vallabh Kallar Channa 1)	SVBPUAT, Meerut	Mutant of C 235	2008	19	135	UP	-
96	JG 14	JNKVV, Jabalpur	(GW 5/7 x P 327) x ICCL 83149	2008	18	-	MP	-
97	JG 6	JNKVV, Jabalpur	(ICCV 10 x K 850) x (h 208 x RS 11	2008	20	-	MP	-

98.	RSG -974 (Abhil-asha)	ARS, Durgapura	K-850 X RSG -515	2008	15	130	Raj.	Suitable for normal and late sown, M.R to dry root rot, wilt, pod borer and root knot nematode
99.	CSJ-140 (Ankur)	ARS, Durgapura	RSG -515 X K -850	2008	16	128	Raj.	Rainfed, Normal & late sown, M.R. to dry root rot, wilt, pod borer & root knot nematode.
100	RSG-959 (Anshul)	ARS, Durgapura	RSG-581X RSG-515	2008	18	130		Normal and Late Sown, M.R. to dry root rot, wilt and root knot nematodes
101	ICPK 2002-29	IIPR, Kanpur	L 144 x H 82-2	2009	21	-	CZ	-
102	PKV Kabuli 4	PDKV Alola	Sel. from local GP	2010	20	-	MS & MP	-
103	IPCK 2004-29	IIPR, Kanpur	[(ICCV 2 x ICCV 88507) x ICCV 42] x ICC 7344	2010	20	-	CZ	-
104	Phule G 0517	MPKV Rahuri	Sel. from local GP	2010	18	-	MS, MP & Karn-ataka	-
105	MNK-1	ARS Gulberga	Sel. from local GP	2010	13	-	MS, MP & Karn-ataka	-
106	Pant Kabuli Chana 1	GBPUA&T, Pantnagar	-	2010	-		Uttra-khand	-
107	Gujarat Juna-garh Gram 3	JAU, Junagarh	ICCV 933001 x ICCV 10	2010	15	-	Gujarat	-
108	HK 05-169	CCS HAU Hisar	HK 92-94 x HK 1	2012	16	-	NEPZ	-
109	JSC 55	RAK COA Sehore	(JAKI 19226 x DCP 20) x JG 412	2012	20	-	CZ	-

110	JSC 56	RAK COA Sehore	(ICCV 91902 x ICCV 10) x ICCV 89230	2012	19	-	CZ	-
111	CSJK -6	ARS Durgapura	-	2012	-		NHZ, NEZ	-

GP- Germ plasm
WR- Wilt resistant
RRR- Root rot resistant
SRA- Suitable for rain-fed areas
MBS- Medium bold seeds
WBS-White and bold seeded
FWR- *Fusarium* wilts resistant
Tolerance.- Tolerance rant
LR- Lodging resistant
WT- Wilt tolerance rant
CRR- Collar rot resistant
SS- Small seeded
BS- Bold seeded
ABR- *Aschochyta* blight resistant
SR- Stunt resistant
Res.- Resistant

DONORS IDENTIFIED AT ARS DURGAPURA, JAIPUR

1. RSG-143-1 Drought tolerant
2. RSG-540 Long pod
3. RSG-581 Dry root rot and wilt tolerant
4. RSG-669 Profuse podding from base
5. CSJ-160 Extra early
6. RSG-895 White flower, early and double poded
7. RSG-974 BGM and dry root rot tolerant
8. RSG-902 Dry root rot tolerant, double poded
9. RSG-931 Dry Root Rot tolerant, double poded
10. CSJD-916 Tolerant to multiple diseases, nematode and drought, high nodulation
11. CSJ-138 High protein content
12. CSJ-146 Nematode tolerant against all species

13. RSG-991	Green seeded and value addition, Drought tolerant, dry root rot and Wilt tolerant
14. RSG-865	Bold seeded, tolerant to dry root rot
15. RSG-965	Medium bold, tolerant to dry root rot

REFERENCES

Annual Report AICRP on Chickpea. 2009. AICRP on Chickpea, IIPR, Kanpur.

Ali, Masood and Mishra, J.P. 2004. *Rabi* pulses (Chickpea, lentil, lathyrus and frenchbean). Text book of field crops production. Directorate of information and publication of Agriculture by Rajendra Kumar. ICAR Krishi Anusandhan Bhavan, Pusa, New Delhi. **PP** : 317-371.

Bahl, P.N. 1988. Breeding approaches for self pollinated crops "Grain legumes". Plant Breeding by V.L. Chopra. Oxford and IBH Publishing Co. Pvt. Ltd. New Delhi, Mumbai and Kolkata. **PP:** 121-137.

Janoria, M.P.; Gour, V.K. and Singh, C.B. 1984. Perspectives in grain legumes. National Seminar on "New Dimensions in Pulse Research and Development" was held at Jabalpur.

Khedar, O.P. and Maloo, S.R. 2012. Seed production in pulses. Pulses production Strategies. by Dr. P. Murugesa Bhoopati and Dr. M. Paramathma. Sri Shakthi Promotional Litho Process. S.F. No. 283 Masaniamman Nagar, Anna Nagar East, Edarpalayam, Coimbatore- 641025. **PP:** 347-362.

Khedar, O.P.; Maloo, S.R. and Punia, S.S. 2012. Pulses in Rajasthan: An overview. Pulses production Strategies. by Dr. P. Murugesa Bhoopati and Dr. M. Paramathma. Sri Shakthi Promotional Litho Process. S.F. No. 283 Masaniamman Nagar, Anna Nagar East, Edarpalayam, Coimbatore- 641025. **PP:** 191-226.

Khedar, O.P.; Singh, R.V.; Shrimali, Mahesh and Singh, N.P. 2008. Pulses: Status and Cultivation Technology. Aaviskar Publishers, Distributors, Jaipur (Raj.)

Mishra, S.P. 2004. Project Coordinator's Report. All India Coordinated Research Project on Chickpea. Indian Institute of Pulses

Research, Kanpur. Annual Group meet was held at JAU, Junagarh, Sept. 15-17, 2004.

Singh, N.P. and Sewak, Shiv 2007. Recent achievements in chickpea research. National Symposium on Legumes for Ecological Sustainability: Emerging Challenges and Opportunities was held at IIPR, Kanpur, Nov. 3-5, 2007 **PP:** 49-55.

Singh, N.P.; Sewak, Shiv; Iquebal, M.A.; Chaturvedi, S.K. and Nath, Omkar 2009. Improved varieties of chickpea in India. AICRP on Chickpea, IIPR, Kanpur.

Singh, N.P.; Sewak, Shiv and Iquebal, Mir Asif 2009. All India Coordinated Research Project on Chickpea: A Profile IIPR, Kanpur.

4

Physiology of Drought Tolerance in Chickpea

S.C. Gupta

INTRODUCTION

Plants grow and develop through a well defined process of growth and differentiation requiring an optimum supply of water, minerals, oxygen, carbon dioxide, suitable temperature, light etc. Deviation in any of these abiotic factors adversely affects the crop productivity. Scarcity or excess supply of any of these abiotic factors limits the normal functioning of plant processes like germination, seedling growth, vegetative growth, leaf area development, dry matter accumulation, flowering and fruiting all of which are sensitive to environmental conditions. Any unusual, natural or anthropogenic aberration in the environment, thus affects the plant growth, development and productivity.

The estimates show that world over the arable land is about 1474 million hectares out of which the un-irrigated area is tremendous (about 1247 million hectares, see: Kumar 2001). In this un-irrigated area drought is an important abiotic factor limiting the crop production. The drought can be primary stress produced by water deficit in the environment surrounding the plant and/or a secondary stress induced by chilling, freezing, heat or salt stress

(Levitt 1980). In India about 36% of the land area is arid or semi-arid while the rest 64% is exposed to temporary drought during the crop season.

Water is the most important and essential ingredient required for various plant processes during the life cycle of a plant. The important functions of water in plants are:

1. It gives turgor to plant cells, essential to promote cell enlargement.
2. It is a raw material for photosynthesis.
3. It helps in monitoring plant temperature through transpirational cooling of plant surfaces.
4. It acts as a solvent and provides a medium for chemical reactions.
5. It provides medium for the transport of inorganic and organic solutes.
6. It causes hydration and neutralization of charges on colloidal molecules. For enzymes, water of hydration helps to maintain structure and facilitates catalytic functions.

Thus, water is an inevitable ingredient in plant life and any aberration in its availability due to drought will affect all the plant processes and in turn the growth and yield of crops.

DROUGHT STRESS

The availability of water during the whole plant life is determined not only by rainfall alone, but by the soil characteristics and evaporative demand also. Hence, the intensity and duration of drought become highly location specific. Scarcity of available water in plants results into a situation well known as drought. The erratic pattern of rainfall distribution further adds to the uncertainty of stage specificity to the drought. Thus a working definition of drought would be "the inadequacy of water availability, including precipitation and soil moisture storage capacity, in quantity and distribution, during the life cycle of a crop to restrict the expression of its full genetic yield potential". Generally drought manifests itself when various combinations of physical and environmental factors cause internal water deficit in crops sufficient to reduce their productivity. Arnon (1975) differentiated between "atmospheric drought" and "soil drought", because the response of plants to these two types of droughts may be different.

A. Atmospheric drought

Low air humidity, frequently accompanied by hot, dry winds results into atmospheric drought. It may occur even under conditions of relatively high soil moisture. This type of drought may have sudden occurrence.

B. Soil drought

When the availability of soil moisture to the plant is less than evapotranspiration, it results in soil drought. It is usually gradual and progressive.

Pulses often experience two types of drought:

1. Intermittent drought- may occur during any of the growth stages due to soil moisture deficit.
2. Terminal drought – may occur during crop maturity stage due to receding soil moisture, more crop demand and increasing air temperatures.

EFFECTS OF DROUGHT

Scarcity of water adversely affects the plant growth (Ali *et al.* 1998) at different levels which can be summarized as follows:

A. Morphological effects: At whole plant level

1. Reduction in dry matter accumulation (Fig. 4.1)

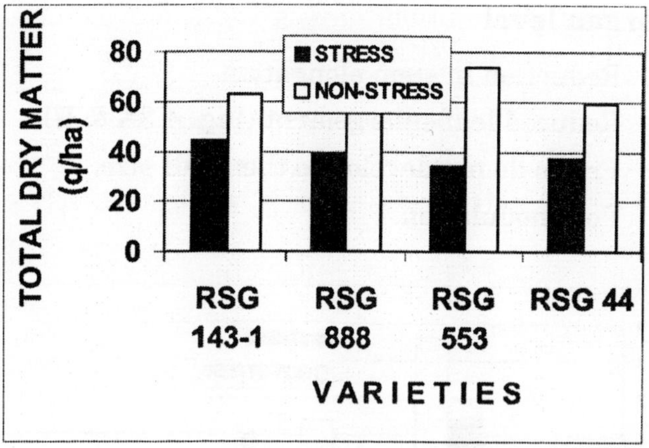

Fig. 4.1. Effect of moisture stress on above ground total dry matter production in some chickpea genotypes as compared to the non-stress conditions in fields

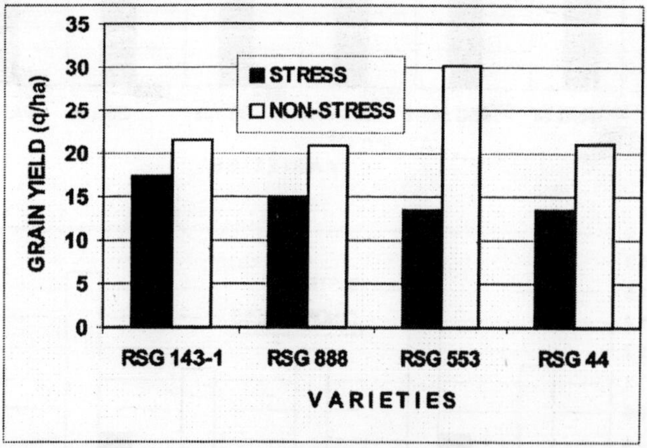

Fig. 4.2. Effect of moisture stress on grain productivity in some chickpea genotypes as compared to the non-stress conditions in fields

2. Reduction in seedling vigour and poor crop establishment (Singh 1990).
3. The whole plant stature is reduced.
4. Crop productivity is reduced (Fig. 4.2).
5. The crop survival becomes difficult under severe water deficit.

B. At organ level

1. Reduction in stem elongation.
2. Reduced leaf enlargement (Fig. 4.3A & B).
3. Fruits do not develop to their full size.
4. Poor nodulation.

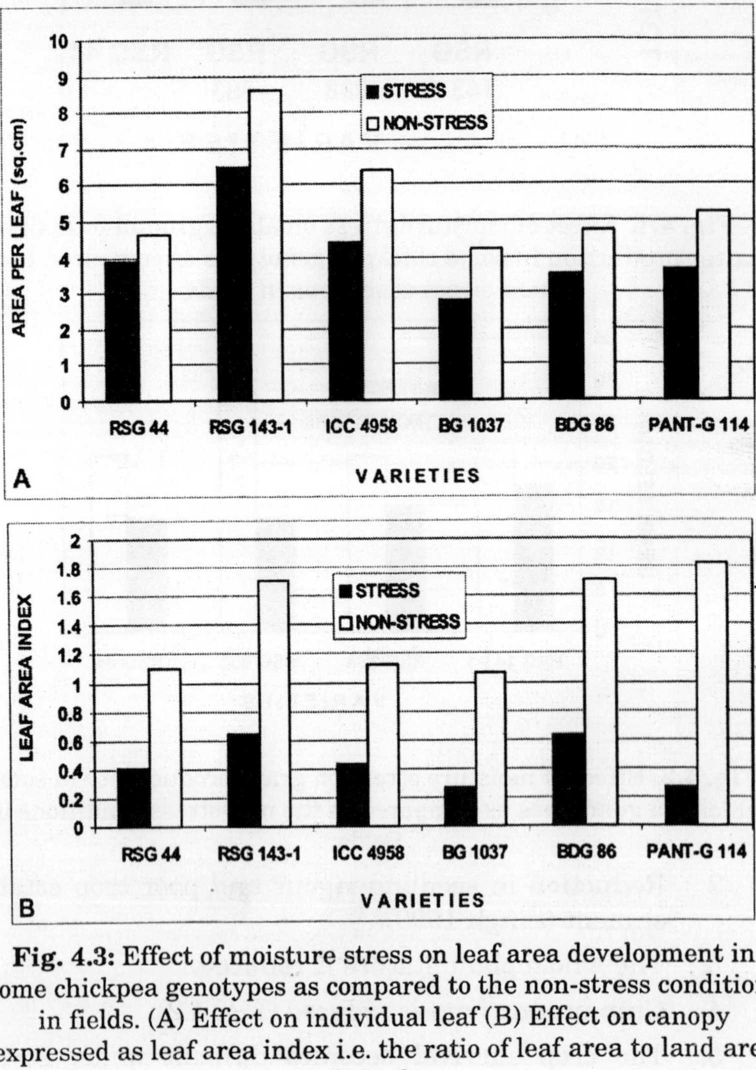

Fig. 4.3: Effect of moisture stress on leaf area development in some chickpea genotypes as compared to the non-stress conditions in fields. (A) Effect on individual leaf (B) Effect on canopy expressed as leaf area index i.e. the ratio of leaf area to land area in a plot

C. At cellular level

1. Poor cell divisions.
2. Reduction in cell enlargement.
3. Poor cell differentiation.

D. Physiological effects

Reduction in

1. Mineral uptake due to air filled pores between the soil particles and there being no direct pathway from the soil to the root surface (Nye and Tinker, 1977; see: Pugnaire *et al.* 1999)
2. Translocation of solutes
3. Plant water potential (Garg *et al.* 2001)
4. Stomatal conductance
5. Carbon dioxide diffusion, thus adversely affecting photo-synthesis (Fig. 4.4)
6. Nitrate reductase activity (Garg *et al.* 2001)
7. Chlorophyll contents (Garg *et al.* 2004)
8. Specific leaf area (Fig. 4.5)

Increase in –

1. Hydrolysis of starch to sugars which in turn may adversely affect the photosynthetic carbon fixation (Garg *et al.* 2001).
2. Respiratory break down of reserve solutes.
3. Peroxidase activities (Singh and Kumar 1992).
4. Production of reactive oxygen forms ($O_2^{\cdot-}$, superoxide; Hydroxyl radicals; Hydrogen peroxide) which cause damage to the enzymes, the chloroplast pigments, the membrane lipids and proteins, and alterations of their structural integrity (Anderson *et al.* 1990; Navari–Izzo *et al.* 1994),

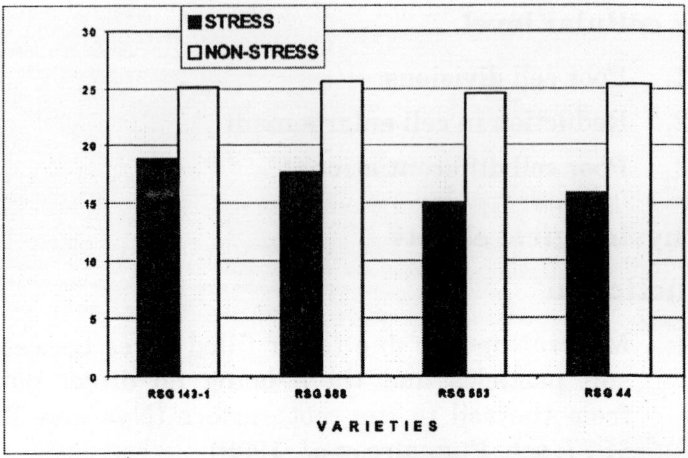

Fig. 4.4: Effect of moisture stress on photosynthetic rate in some chickpea genotypes as compared to the non-stress conditions in fields. See that the tolerant genotype RSG 143-1 maintains comparatively higher photosynthetic rate under stress

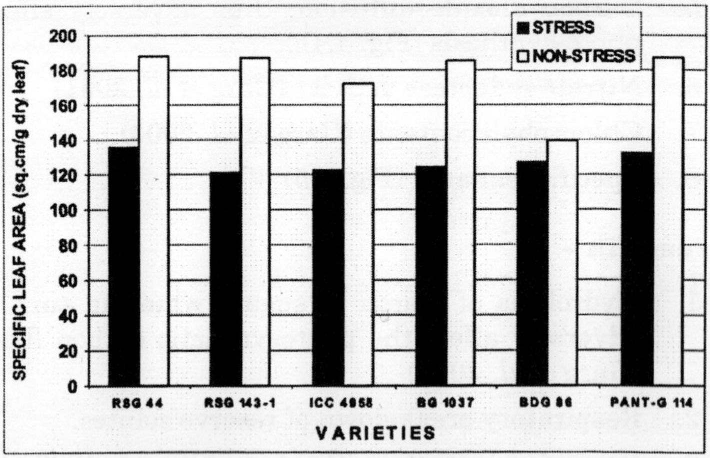

Fig. 4.5: Effect of moisture stress on specific leaf area i.e. the ratio of leaf area to its dry weight in some chickpea genotypes as compared to the non-stress conditions in fields

5. ABA level in roots, which is translocated to leaves and results in increased stomatal resistance (Zhang *et al.* 1987),

6. Ethylene production, which results in early abscission of leaves, flowers and fruits (Taiz and Zeiger 1998),
7. Proteolytic activities, with increased protein hydrolysis it result into accumulation of toxic products like ammonia, which may cause tissue injuries (Taiz and Zieger 1998),
8. Soluble osmolytes like proline, sugars, etc. (Gupta *et al.* 2000, Ghorbanli *et al.* 2003; Fig. 4.6),
9. Canopy temperature due to reduced transpiration (Fig. 4.7).

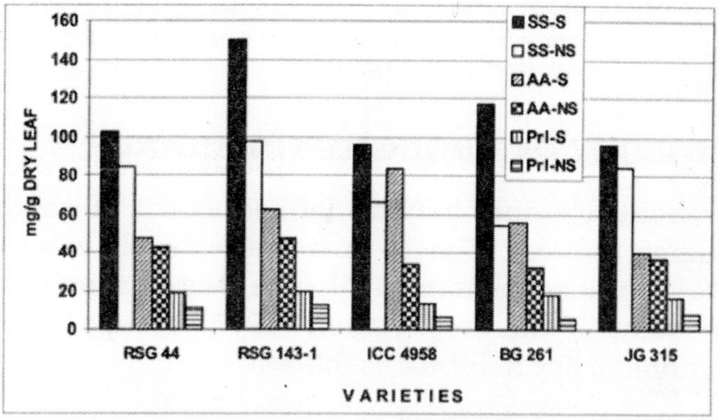

Fig. 4.6: Effects of moisture stress on accumulation of osmolytes in leaf tissue in some chickpea genotypes as compared to the non-stress conditions in fields. See that the accumulation of osmolytes under stress is comparatively higher in the tolerant genotype RSG 143-1 (SS = Soluble Sugar, AA = Soluble Amino Acids, Prl = Proline, -S = Stress, -NS = Non-Stress; the values of Amino acids and praline in the figure are ten times the original values)

Thus, the metabolic activities including nucleic acid and protein synthesis are impaired under drought stress. The increased respiration combined with reduced photosynthesis may cause starvation (Björkman *et al.* 1980). Plants are usually more sensitive to the stress during late vegetative and reproductive stages because of the competition for assimilates between roots and developing flowers and fruits (Silim and Saxena, 1993).

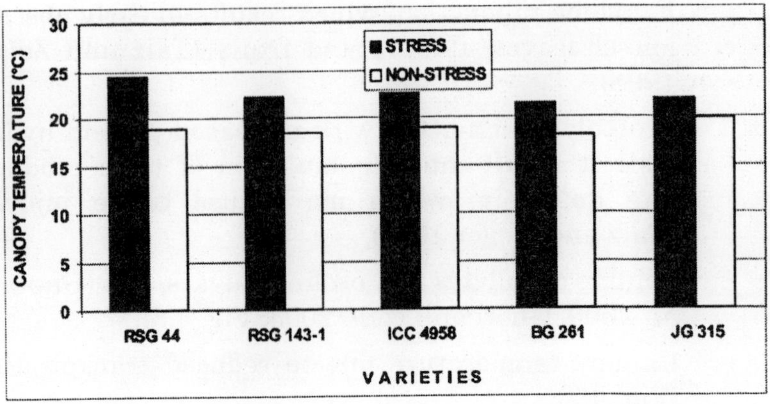

Fig. 4.7: Effect of moisture stress on canopy temperature in some chickpea genotypes as compared to the non-stress conditions in fields

PHYSIOLOGY OF DROUGHT TOLERANCE

Drought tolerance may be defined as the mechanism (s) minimizing the loss of yield in a water deficit environment relative to the maximum yield in a water constraint free management of the crop. Arnon (1975) grouped the plant adaptations to dry habitats as follows:

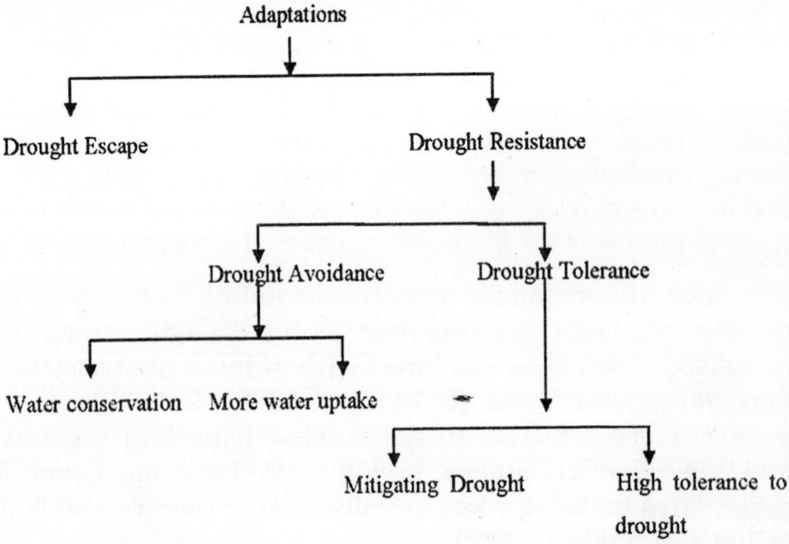

Kramer (1980) also proposed the mechanisms that operate in different plant systems which were further elaborated by Turner (1986) and Blum (1988). According to them, the three drought resistance categories are:

A. Drought escape

B. Dehydration postponement

C. Dehydration tolerance

A. Drought Escape

It is an adaptation of matching the phenology to the water supply and allows completing the plant's life cycle during favorable environment before the commencement of drought stress. In chickpea, such adaptation has been of limited use because in response to water deficit the plants show early flowering and forced maturity and it is at the cost of plant productivity. There has been some success through selections for early flowering under environments prone to terminal drought (Hall *et al.* 1978; Hall and Patel 1985; Subbarao *et al.* 1995). At International Crops Research Institute for Semi Arid Tropics, Hyderabad, India, some short duration chickpea lines like ICCV-2, ICCC-37, ICCV-10 and KAK-2 have been developed which effectively escape end-of-season drought (Dar *et al.* 2001; Serraj *et al.* 2003). Also, phenological plasticity has been considered an important characteristic in drought prone environments (Turner, personal communications). However, the problem of unpredictable and intermittent drought cannot find a solution in early flowering types.

B. Dehydration postponement

It is the mechanism of maintaining turgor during drought period through efficient water uptake and reduction in water loss through osmotic adjustment. The plant water indices like water retention and water content in fresh samples have been suggested as screening tools for tolerance to drought in chickpea (Pannu *et al.* 1993, Gupta and Sharma 2006).

Efficient water uptake is largely a function of rooting system of the plant for increasing soil water extraction. There have been reports of genotypic variations in rooting depth and its ability to extract water at different soil profiles having impact on seed yield (Gregory 1988; Mathews *et al.* 1988; White and Castillo 1990; Wright *et al.* 1991). Supporting the earlier findings (Gupta *et al.* 1987), Saxena *et al.* (1990) has reported that denser and longer root system in chickpea is advantageous under drought stressed environments. The chickpea genotypes ICC-4958 and FLIP 87-59°C have been identified as drought tolerant lines with higher root biomass (Saxena *et al.*1994; Singh *et al.* 1996) and are presently being used as donor parents for this trait. Another genotype showing drought tolerance is RSG 143-1 (Gupta *et al.* 2003), which is having more root biomass in the deeper soil depths (>60 cm), which results in extracting more moisture from the deeper soil profiles (Fig. 4.8).

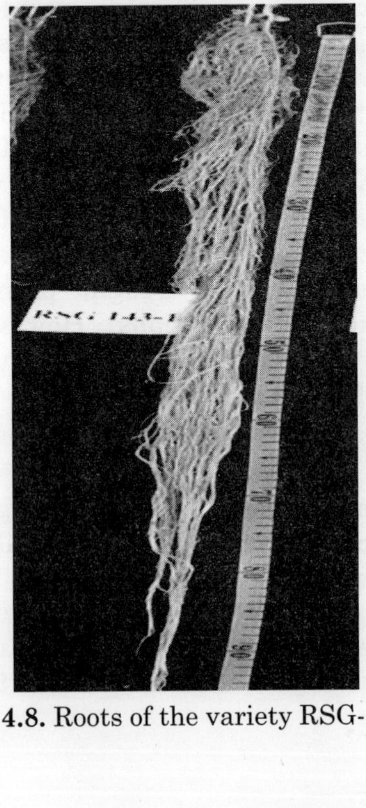

Fig. 4.8. Roots of the variety RSG-143-1

Several techniques have been used to screen the genotypes for their rooting characteristics. The technique of Robertson *et al.* (1985) of introducing selective herbicides at depth has been used by Khalfaoui and Havard (1993) for identifying the deep rooting lines. Sand culture technique for the similar purpose in chickpea has been used by Saxena *et al.* (1994) and Krishnamurthy *et al.* (1996). Other techniques like aeroponics, root pulling resistance, tensiometer etc. have also been used for screening the varieties (Subbarao *et al.* 1995, Mia *et al.* 1996a, 1996b). In rice the molecular markers have been identified to characterize the root morphology (Champoux *et al.* 1995; Ray *et al.* 1996). Similar studies in grain legumes will be of immense help in screening a large number of genotypes for their rooting behavior.

There have been debates (Passioura 1994) whether the higher root biomass will be a beneficial characteristic (Jordan *et al.* 1983), or the reduced underground growth (Passioura 1983) will increase seed yield under water stress conditions. The work at ICRISAT (Saxena *et al.* 1994; Krishnamurthy *et al.* 2003) suggests that the genotypes with higher root biomass perform better under limited moisture.

Reduction in water loss is mainly accomplished through regulation of stomatal closure under water stress conditions. The moisture status of leaves and the ambient humidity have the major influence on stomatal conductance (Turner 1991). Variations in stomatal conductance in response to varying leaf water potential have been reported in grain legumes (Lawn 1982; Muchow 1985, Flower and Ludlow 1986). There are however, reports of decreasing stomatal conductance in response to drying soils, even when leaf hydration is maintained (Blackman and Davies 1985; Gollan *et al.* 1986). Thus, a stimulus may be transmitted to the leaves which control the closure of stomata. Abscisic acid (ABA) has been reported to provide such kind of stimulus in plants (Henson *et al.* 1989; Turner and Henson 1989; Davies *et al.* 1[illegible]

Stomatal conductance controls not only the water loss by the plant but also the rate of photosynthesis and regulates the gaseous exchange. Therefore, any reduction in conductance will reduce both the water loss as well as the carbon fixation by the plant. In chickpea the tolerant genotypes make a compromise where by the conductance is reduced to decrease the water loss to the extent that stomata remain partially open to allow sufficient carbon dioxide entry into the leaves and a lesser reduction in photosynthetic rate is observed (Gupta *et al.* 2000).

Variations in the ABA production ability among the genotypes occur in response to water stress. However, it has been reported that the genes for ABA production under drought are located on chromosomes other than the chromosomes bearing the genes contributing to yield, which very clearly suggested that the drought tolerance in terms of higher ABA production in response to water stress may not have high correlation with grain productivity (Quarrie *et al.* 1995). Hence, the way-out to develop drought tolerance would be to explore the genetic variability for ABA production and introducing this trait into a higher yielding cultivar.

Leaflet closure has been shown to reduce water loss in groundnut (Ramesh Babu and Rao 1983). Such adaptation protects the plants from exposure to excessive solar radiations and thus provides a mechanism to tolerate the stress. Similar reduction in leaflet angle has been observed in chickpea in response to moisture stress (Choudhary 2002).

Osmotic adjustment is the active accumulation of solutes in plant cells in response to water deficit to maintain turgor or to reduce the rate of turgor loss, with a decrease in water potential (Turner and Jones 1980; Morgan 1984). It involves the active accumulation of solutes, such as sugars, amino acids, organic acids, NO_3^-, Cl^-, and K^+ in response to drought which lowers osmotic potential and maintains cell turgor at lower soil water potential (Khanna Chopra *et al.* 1995). Osmotic adjustment has been shown to be beneficial in imparting drought tolerance (Wery *et al.* 1993;

Choudhary 2002) at low water potential, in maintaining stomatal conductance and photosynthetic rate (Jones and Rawson 1979; Ludlow 1980, 1987), in delaying leaf senescence (Hsiao *et al.* 1984; Flower and Ludlow 1986, 1987), in reducing flower abortion (Morgan and King 1984) and in enhancing root growth and water uptake (Morgan and Condon, 1986). In chickpea, more lowering of leaf osmotic potential is shown by the tolerant genotypes (Gupta *et al.* 2000). Higher leaf moisture retention ability has also been observed in such genotypes in our laboratory and elsewhere as well (Ali *et al.* 1998). An osmotic adjustment of 0.1 to 1.6 MPa has been reported in different grain legumes (see: Subbarao *et al.* 1995). In chickpea, Morgan *et al.* (1991) reported an osmotic adjustment of 0.6 MPa in leaves, and showed to be correlated with yield under drought conditions. Flower and Ludlow (1986, 1987) explained that in grain legumes the osmotic adjustment delays leaf senescence and increases the remobilization of food reserves, thus enhances yield under water limited conditions.

Genotypic variability has also been found to occur in maintaining canopy temperature under stress conditions. Chauhan and Singh (2003) have proposed that lower canopy temperature under water stress might be attributed to higher water use efficiency and osmo-regulation in the tolerant chickpea genotypes.

There have been genetic studies identifying the trait loci for osmotic adjustment, mostly in cereals (Lilley *et al.* 1996; Morgan *et al.* 1986). Identification of the molecular markers like the proteins, dehydrins, synthesized in response to water deficit (Bartels *et al.* 1993) will simplify the selection of genotypes for complex morphological, physiological and biochemical traits from a large population. Also new tolerant genotypes can be developed through DNA transfer technology application in grain legumes (Christou 1994).

C. Dehydration tolerance

It is the ability of plants to maintain cellular metabolism at low water potential. Generally, when the water deficit or plant dehydr-ation reaches a critical level (for many legumes leaf water potential -1.8 to -8.2 MPa, Flower and Ludlow 1987), the plants die due to disordered membrane functions. The membrane fluidity for ion leakage has been used to select out the plants having more stable membranes and thus tolerance to high temperature and water deficit (Srinivasan *et al.* 1996; Deshmukh and Kushwaha 2002). In chickpea, the genotypes showing tolerance to drought have lower membrane injury in response to heat shock treatment (Gupta *et al.* 2000, Fig. 4.9). Wery *et al.* (1993) have also emphasized on high membrane stability under stress for tolerance to drought.

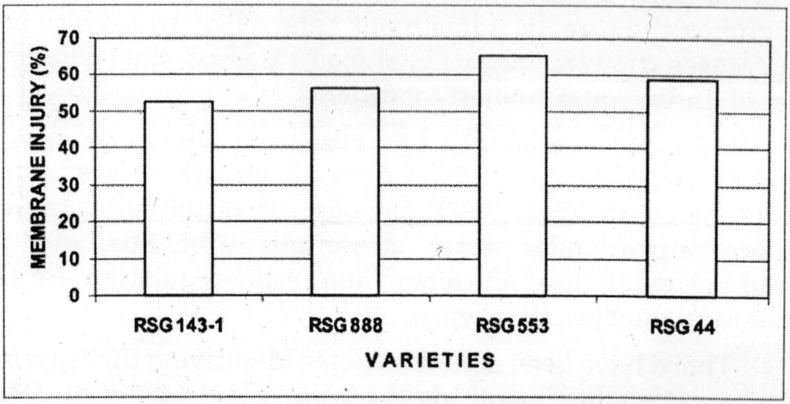

Fig. 4.9. Variations in leaf membrane injury due to stress in some chickpea genotypes. See that the tolerant genotype RSG 143-1 shows comparatively the lowest membrane injury

However, there are reports that no relationship exists between electrolyte leakage from dehydrated leaves and crop performance under drought (Blum, 1988) and hence the technique is not in much use in the breeding programmes. But, looking to the active role of membranes in plant metabolism, one can expect the importance of membrane integrity under stress conditions and thus more work is required to quantify this characteristic for use as a selection criterion for drought tolerance. Sampling of leaves for measurement of ion leakage is important for uniformity of results and their valid comparisons. For example in chickpea, the 5th, 6th & 7th fully expanded healthy leaves of the main shoot from the top can serve as good samples for physiological analyses.

The tolerance of grain legumes to the lowest water potential experienced by the viable leaves has been reported to vary from -8.2 to -1.8 MPa (Shackel and Hall 1983; Flower and Ludlow 1986, 1987; Sinclair and Ludlow, 1986). However, osmotic adjustment has been reported to play an important role in dehydration tolerance (Hsiao *et al.* 1984) but, the factors other than osmotic adjustment cannot be ruled out as there existed a poor correlation between osmotic adjustment and the lethal osmotic potential for the leaves (Lilley *et al.* 1996). More lowering of osmotic potential in response to water stress has been reported in the drought tolerant chickpea genotypes (Gupta *et al.* 2000, Fig. 4.10) and has been associated with mitigation of the deleterious effects of moisture stress (Turner and Begg 1981; Morgan 1984). Accumulation of cellular osmolytes like soluble sugars, free amino acids, proline, glycine betaine etc. has been shown to occur as a general feature under moisture stress conditions (Gupta *et al.* 1994, 2000; Jha and Singh 1997). Out of these osmolytes, proline has been amply implicated to provide an energy source for utilization during plant recovery from stress (Aspinall and Paleg 1981; Paleg *et al.* 1981). The physiological significance of proline accumulation has been debatable as in some cases it has been reported to be an stress injury effect (Hanson *et al.* 1977) while in others, a protective agent of enzymes and cellular structures (Schobert and Tschesche 1978) and still in other cases as a storage compound which is used up during rapid growth after the stress is released (Barnett and Naylor 1966). In chickpea it has been reported that proline accumulation increases under water stress and the accumulation is more in drought tolerant genotypes (Gupta *et al.*, 2000).

PEG-6000 induced tolerance to drought in callus cell lines has also been reported as due to higher levels of K^+ and proline (Gulati and Jaiwal 1994).

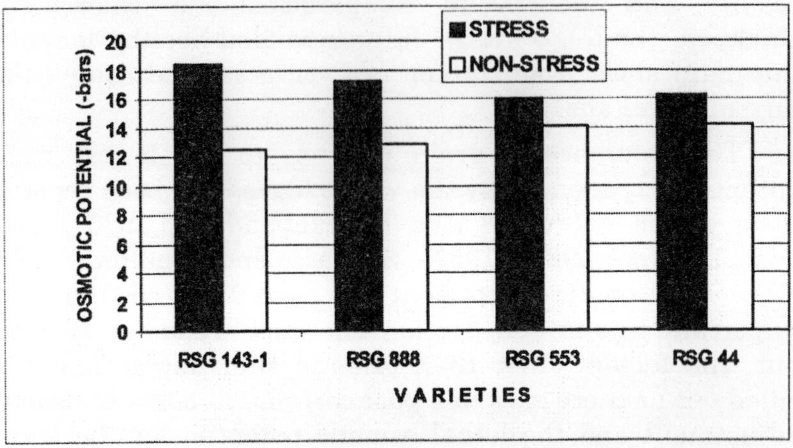

Fig. 4.10: Effect of moisture stress on leaf osmotic potential of some chickpea genotypes as compared to the non-stress conditions in fields. See that the tolerant genotype RSG 143-1 shows comparatively the lowest osmotic potential under stress

Silim and Saxena (1993) in their studies on chickpea reported that deep root system, high leaf water potential at dawn and large number of seeds per plant were associated with drought tolerance. Early seedling vigour has also been shown to play some role for drought tolerance under moisture stress. Harris *et al.* (1999, 2001) found that 10 hour seed soaking in water prior to sowing in fields resulted in faster emergence, better stands, more vigorous plants, better drought tolerance, earlier flowering, earlier harvest and higher grain yield in chickpea. In our experiments under moisture stress conditions, we also found early emergence and more vigourous plants with higher grain yields due to seed priming with water favouring plant performance under moisture stress. About 9% higher seedling vigour indexes were observed in primed seeds (Fig. 4.11).

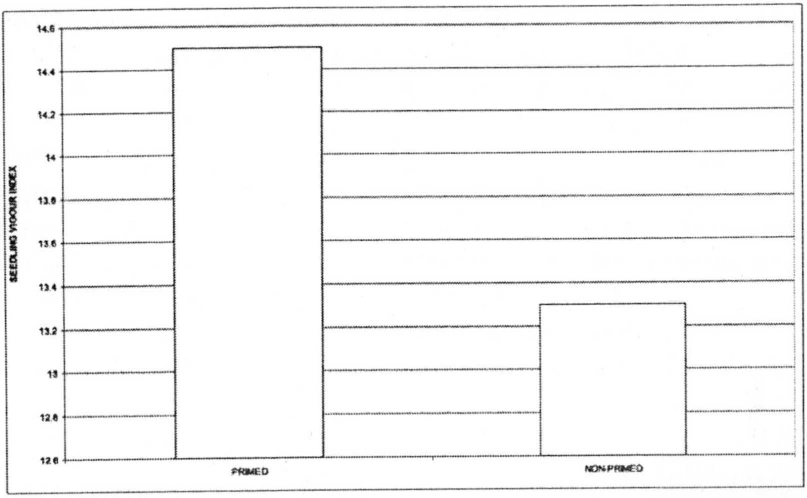

Fig. 4.11. Effect of seed priming prior to sowing on seedling vigour index in chickpea (var. RSG 888) as compared to non-primed seeds

Remobilization of carbon and nitrogen from vegetative parts to developing seed is important in chickpea because the pod filling occurs when net photosynthesis and nitrogen fixation falls due to the onset of terminal drought (Davies *et al.* 2000). Desi genotypes show better drought tolerance, partly due to their ability of better remobilization and higher pod number.

The various physiological adjustments in response to moisture stress allows a tolerant chickpea genotype to maintain lower drought susceptibility index (DSI, Gupta *et al.* 2000), which is a good measure for drought tolerance in chickpea. Lower DSI value signifies that the genotype has yield sustainability under the defined drought stress conditions.

POTENTIAL AND LIMITATIONS OF USING PHYSIOLOGICAL BASE FOR DEVELOPMENT OF DROUGHT TOLERANCE

Physiological traits critically explain how a plant system economically manages the water use and the metabolic processes under water stress conditions. Up to a certain limit such economic water use may not affect the grain production appreciably but under severe water stress, survival of life naturally becomes more important for physiological processes, which on priority manage to protect the life of the plant irrespective to the loss in grain yield. Struggle for the survival is a natural phenomenon while production oriented crop manipulations are all anthropogenic fighting against the rule of nature. Hence, significant correlations are expected to exist among the physiological traits involved in water economy management, and not with grain yield under drought conditions, which in true sense signifies the drought tolerance of a system. However, incorporation of the physiological traits responsible for drought tolerance into a high yielding genotype with good agronomic background may be a feasible solution to overcome the problem of water stress and sustainable crop production. Mutation breeding has also shown some success in developing the tolerant systems (Ahulwalia *et al.* 2004). The specific genes encoding for enzymes required for the biosynthesis of various osmo protectants or modifying membrane lipids etc. are being identified and efforts are in progress to transfer the same in high yielding cultivars (Sharma and Lavanya 2002). Alfin 1 has proved to be an essential gene for root growth and in transgenic plants its over expression confers a many fold increase in root growth, an advantageous trait for stress conditions (Winicov 2000).

REFERENCES

Ahulwalia, B.S., Maluszynski, M. & Nichterlein, K. 2004. Global impact of mutation derived varieties. *Euphytica.* 135(2): 187-204.

Ali, Y., Haq, M. A., Alam, S.S. & Ahmad, N. 1998. Variation in drought tolerance in ten chickpea varieties. II: Studies on some physiological characteristics and their relation to growth. *Pakistani Journal of Biological Sciences*. 1(2): 109-110.

Anderson, J.V., Hess, J.L. & Cheione, B.J. 1990. Purification, characterization and immunological properties for two isoforms of glutathione reductase from eastern white pine needles. *Plant Physiol*. 94: 1402-1409.

Arnon, I. 1975. Physiological principles of dry land crop production. In: *Physiological aspects of dry land farming* (Ed. U.S. Gupta). Oxford and IBH Publishing Co. **PP:** 3-145.

Aspinal, D. & Paleg, L.G. 1981. Proline accumulation: Physiological aspects. In: The physiology and Biochemistry of Drought Resistance in Plants (Eds. L.G. Paleg and D. Aspinal), Sydney, *Acadmic Press*. **PP:** 205-241.

Barnett, N.N. & Naylor, A.N. 1966. Amino acid and protein metabolism in Bermuda grass during water stress. *Plant Physiol.* 41: 1222-1230.

Bartels, D., Alexander, R., Schneider, K., Elster, R., Velasco, R., Alamillo, J., Bianchi, G., Nelson, D. & Salamini, F. 1993. Desiccation related gene products analyzed in a resurrection plant and in barley embryos. In: Plant Responses to Cellular Dehydration during Environmental Stress (Eds. T.J. Close, E.A. Bray & Rockville, M.D.) *American Society of Plant Physiologists*. **PP:** 119-127.

Björkman, O., Badger, M. R. & Armond, P. A. 1980. Response and adaptation of photosynthesis to high temperatures. In: Adaptation of plants to water and high temperature stress (Eds. N.C. Turner & P.J. Kramer). Wiley, New York. pp. 233-249.

Blackman, P.G. & Davies, W.J. 1985. Root to shoot communication in maize plants – the effects of soil drying. *J. Exp. Bot.* 36: 39-48.

Blum, A. 1988. Plant breeding for stress environments. *Boca Raton, F.L.:* CRC Press.

Champoux, M.C., Wang, G. Sarkarung, S., MacKill, D.J., O'Toole, J.C., Huang, N. & McCouch, S.R. 1995. Locating genes associated with root morphology and drought avoidance in rice via linkage to molecular markers. *Theoret. Appl. Genet.* 90: 969-981.

Chauhan, S.S. & Singh, D.N. 2003. Genotypic variation among chickpea genotypes in canopy temperature and relationship to grain yield under soil moisture deficit conditions. Abstracts. International chickpea conference, Raipur, Chhattisgarh, India, January 20-22, 2003, **PP:** 76-77.

Choudhary, P. 2002. Physiological and biochemical studies on some drought tolerant and susceptible genotypes of chickpea. Ph.D. Thesis, Deptt. of Botany, University of Rajathan, Jaipur, India.

Christou, P. 1994. The biotechnology of crop legumes. *Euphytica.* 74: 165-185.

Dar, W.D., Kumar, J., Saxena, K.B. Upadhyaya, H.D. & Crouch, J.H. 2001. International Efforts on Improvement of Chickpea and Pigeonpea. In: Souvenir, National Symp. on Pulses for sustainable Agriculture and Nutritional Security, ISPRD, Kanpur. **PP:** 1-7.

Davies, S.L., Turner, N.C., Palta, J.A., Siddique, K.H.M. & Plummer, J.A. 2000. Remobilisation of carbon and nitrogen supports seed filling in chickpea subjected to water deficit. *Australian Journal of Agricultural Research.* 51(7): 855-866.

Davies, W.J., Tardieu, F. & Trejo, C.L. 1994. How do chemical signal work in plants that grow in drying soil. *Plant Physiol.* 104: 309-314.

Deshmukh, P.S. & Kushwaha, S.R. 2002. Variability in membrane injury index in chickpea genotypes. *Indian Journal of Plant Physiol.* 7(3): 285-287.

Flower, D.J. & Ludlow, M.M. 1986. Contribution of osmotic adjustment to the dehydration tolerance of water stressed pigeonpea [*Cajanus cajan* (L.) Millsp.] Leaves. *Plant Cell Environment.* 9: 33-40.

Flower, D.J. & Ludlow, M.M. 1987. Variation among accessions of pigeonpea (*Cajanus cajan*) in osmotic adjustment and dehydration tolerance of leaves. *Field Crops Research.* 17: 229-243.

Garg, B. K., Burman, U. & Kathju, S. 2004. Effect of water stress on moth bean (*Vigna aconitifolia* (Jacq.) Marechal) genotypes. *Indian J. Plant Physiol.* 9(1) :30-36.

Garg, B.K., Kathju, S. & Burman, U. 2001. Influence of water stress on water relations, photosynthetic parameters and nitrogen metabolism of mothbean genotypes. *Biologia Plantarum.* 44(2): 289-292.

Ghorbanli, M., Farzami Sepehr, M. & Farbodnia, T. 2003. The changes in total soluble sugars, starch and protein contents at drought stress in two cultivars of Iranian Cicer (*Cicer arietinum* L.). Abstract. International Chickpea Conference, Raipur, Chhattisgarh, India, January 20-22, 2003. pp. 67.

Gollan, T., Passioura, J.B. & Munns, R. 1986. Soil water status affects the stomatal conductance of fully turgid wheat and sunflower leaves. *Australian Journal of Plant Physiol.* 13: 459-464.

Gregory, P. J. 1988. Root growth of chickpea, fababean, lentil and pea and effects of water and salt stress. In: World Crops: Cool Season Food Legumes (Ed. R.J. Summerfield), Dordrecht, *The Netherlands, Kluwer Academic Publishers.* **PP:** 857-867.

Gulati, A. & Jaiwal, P. K. 1994. *In vitro* selection and characterization of *Vigna radiata* cell lines resistant to PEG induced drought stress. *Acta Physiologiae Plantarum.* 16(1): 53-60.

Gupta, S.C., Bhatia, L.K. & Singh, A.K. 1994. Role of proline and cellular osmotic agents in drought tolerancein cowpea. In: Plant Productivity under Environmental Stress. (Eds. K. Singh and S.S. Purohit). Agro Botanical Publ., Bikaner, India. **PP:** 237-240.

Gupta, S. C., Rathore, A. K., Sharma, S. N. & Saini, R. S. 2000. Response of chickpea cultivars to water stress. *Indian J. Plant Physiol.* 5(3) : 274-276.

Gupta, S. C. & Sharma, S.N. 2006. Leaf moisture retention index (LMRI): An easily measurable physiological character-istic for drought tolerance in chickpea. *Indian Journal Pulses Research.* 19 (2): 83-84.

Gupta, S.C., Sharma, S.N. & Jhabar Singh 2003. Root growth characteristics of some chickpea genotypes under moisture stress. Abstract National Symposium on Improving Crop Production in an Eco-Friendly Environment. Physiological and Molecular Approaches, October 15-17, 2003, Pantnagar, India.

Gupta, S. N., Dahiya, B. S. & Singh, V. P. 1987. Genetic variability for drought tolerance in chickpea. Abstracts. First Symposium on Crop Improvement, 23-27 February, 1987, Ludhiana, India. *Crop Improvement Society of India.* **PP:** 128-129.

Hall, A.E., Foster, K.W. & Waines, J. G. 1978. Crop adaptation to semi-arid environments. In: Agriculture in semi-arid environments (Eds. G. H. Cannell, A. E. Hall & H. Lawton), New York, N.Y., USA, Springer-Verlag. **PP:** 148-179.

Hall, A. E. & Patel, P. N. 1985. Breeding for resistance to drought and heat. In: Cowpea Research, Production and Utilization (Eds. S. R. Singh & K. O. Rachie), Chrichester, USA, John Wiley. **PP:** 137-151.

Hanson, A. D., Nelson, C. E. & Everson, E. H. 1977. Evaluation of free proline accumulation as an index of drought resistance using two contrasting barley cultivars. *Crop Sci.* 17: 720-726.

Harris, D., Joshi, A., Khan, P. A., Gothkar, P. & Sodhi, P. S. 1999. On-farm seed priming in semi-arid agriculture: Development and evaluation in maize, rice and chickpea in India using participatory methods. *Experimental Agriculture.* 35(1): 15-29.

Harris, D., Pathan, A. K., Gothkar, P., Joshi, A., Chivasa, W. & Nyamudeza, P. 2001. On-farm seed priming using participatory methods to revive and refine a key technology. *Agricultural Systems.* 69(1-2): 151-164.

Henson, I. E., Jensen, C. R. & Turner, N. C. 1989. Leaf gas exchange and water relations of lupins and wheat. III. Abscisic acid and drought induced stomatal closure. *Australian Journal of Plant Physiol.* 16: 429-442.

Hsiao, T. C., O'Toole, J. C., Yambao, E. B. & Turner, N. C. 1984. Influence of osmotic adjustment on leaf rolling and tissue death in rice (*Oryza sativa* L.). *Plant Physiol.* 95: 338-341.

Jha, B. N. & Singh, R. N. 1997. Physiological responses of rice varieties to different levels of moisture stress. *Indian J. Plant Physiol.* 2: 81-84.

Jones, M. M & Rawson, H. M. 1979. Influence of the rate of development of leaf water deficits upon photosynthesis, leaf conductance, water use efficiency and osmotic potential in sorghum. *Physiol. Plant.* 45: 103-111.

Jordon, W. R., Dugas, W. A. & Shouse, P. J. 1983. Strategies for crop improvement for drought prone regions. *Agric. Water Manage.* 7: 281 – 299.

Khalfaoui, J. L. B. & Havard, M. 1993. Screening peanut cultivars in the field for root growth: A test by herbicide injection in the soil. *Field Crops Research.* 32:173-179.

Khanna Chopra, R., Moinuddin, Sujata Vasudev, Maheshwari, M., Amita Srivastava, Bahukhandi, D., Vasudev, S. & Srivastava, A. 1995. K^+, osmo regulation and drought tolerance - an overview. Proceedings of the Indian National Science Academi – Part B, *Biological Sciences.* 61(1): 51-56.

Kramer, P. J. 1980. Drought stress and the origin of adaptations. In: Adaptation of plants to water and high temperature stress (Eds. N. C. Turner & P. J. Kramer). New York, N.Y. Academic Press. PP: 7-20.

Krishnamurthy, L., Johansen, C. & Ito, O. 1996. Genotypic variation in root system development and its implications for drought resistance in chickpea. In: Dynamics of Roots and Nitrogen in Cropping Systems of the Semi-Arid Tropics. (Eds. O. Ito, C. Johansen, J. J. Adu-Gyamfi, K. Katayama, J.V.D.K., Kumar Rao & T. J. Rego), Tsukuba, Japan, *Japan International Research Center for Agricultural Sciences*. **PP:** 235-250.

Krishnamurthy, L., Kashiwagi, J. & Serraj, R. 2003. Genetic variations of chickpea (*Cicer arietinum* L.) root traits and their influence on yield stability under drought. Abstracts. International Chickpea Conference, Raipur, Chhattisgarh, India, January 20-22, 2003. **PP:** 87.

Kumar, D. 2001. Breeding approaches in arid legumes for abiotic stresses. In: Pulses for sustainable agriculture and nutritional security (Eds. Masood Ali, S.K. Chaturvedi & S.N.Gurha) - A Proceeding of National Symposium, ISPRD, IIPR, Kanpur. pp. 73-81.

Lawn, R. J. 1982. Responses of four grain legumes to water stress in south eastern Queensland. I. Physiological response mechanisms. *Australian Journal of Agricultural Research.* 33:481-496.

Levitt, J. 1980. Response of plants to environmental stresses. Vol. 1, 2nd Ed. Academic Press, New York.

Lilley, J. M., Ludlow, M. M., McCouch, S. R. & O'Toole, J. C. 1996. Locating QTL for osmotic adjustment and dehydration tolerance in rice. *J. Exp. Bot.* 47: 1427-1436.

Ludlow, M.M. 1980. Adaptive significance of stomatal responses to water stress. In: Adaptation of Plants to water and High Temperature Stress (Eds. N. C. Turner & P. J. Kramer), New York, N.Y., John Wiley Publ. **PP:** 123-128.

Ludlow, M.M. 1987. Contribution of osmotic adjustment to the maintenance of photosynthesis during water stress. *Progress in Photosynthesis Research* (Ed. J. Biggins) Vol. 4, Dordrecht, The Netherlands, Martinus Nijhoff Publ. **PP:** 161-168.

Mathews, R.B., Harris, D., Rao, R.C.N., Williams, J.H. & Wadia, K.D.R. 1988. The physiological basis for yield differences between four genotypes of groundnut (*Arachis hypogaea*) in response to drought. I. Dry matter production and water use. *Experimental Agriculture.* 24: 191-202.

Mia, M.W., Yamanchi, A. & Kono, Y. 1996a. Root system structure of six food legume species: Inter- and intraspecific variations. *Jpn. J. Crop. Sci.* 65: 131-140.

Mia, M.W., Yamanchi, A. & Kono, Y. 1996b. Plasticity of tap root elongation growth of several food legume species. *Jpn. J. Crop. Sci.* 65: 368-378.

Morgan, J.M. 1984. Osmoregulation and water stress in higher plants. *Annual. Rev. Plant Physiol.* 35: 299-319.

Morgan, J.M. & Condon, A.G. 1986. Water use, grain yield and osmoregulation in wheat. *Australian Journal of Plant Physiol.* 13: 523-532.

Morgan, J.M. & King, R.W. 1984. Association between loss of leaf turgor, abscisic acid levels and seed set in two wheat cultivars. *Australian Journal of Plant Physiol.* 11: 143-150.

Morgan, J.M., Hare, R.A. & Fletcher, R.J. 1986. Genetic variation in osmoregulation in bread and durum wheat and its relationship to grain yield of field environments. *Australian Journal of Agricultural Research.* 37: 449-457.

Morgan, J. M., Rodriguez-Maribona, B. & Knights, E.J. 1991. Adaptation to water deficit in chickpea breeding lines by osmoregulation: Relationship to grain–yields in the field. *Field Crops Research* 27**:** 61-70.

Muchow, R. C. 1985. Stomatal behaviour in grain legumes grown under different soil water regimes in a semi-arid tropical environment. *Field Crops Research* 11: 291-307.

Navari-Izzo, F., Pinzino, C., Quartacci, M.F., Sgherri, C.L.M. & Izzo, R. 1994. Intracellular membranes: Kinetics of superoxide production and changes in thylakoids of resurrection plants upon dehydration and rehydration. *Proc. R. Soc. Edinburgh.* 102 B: 187-191.

Nye, P.H. & Tinker, P.B.H. 1977. Solute movement in the soil-root system. Oxford, U.K., Blackwell.

Paleg, L.G., Douglas, T.J., Van Daala & Kecch, B.D. 1981. Proline, betaine and other organic solutes protect enzymes against heat inactivation. *Australian Journal of Plant Physiol.* 8: 107-114.

Pannu, R.K., Singh, D.P., Singh, P., Choudhary, B.D. & Singh, V.P. 1993. Evaluation of various plant water indices for screening the genotypes of chickpea under limited water environment. *Haryana Journal of Agronomy.* 9(1): 16-22.

Passioura, J.B. 1994. The yield of crops in relation to drought. In: Physiology and determination of crop yield (Eds. K.J. Boote,

J.M. Bennett, T.R. Sinclair & G.M. Paulsen), Madison, WI: Crop Science Society of America, **PP:** 343-359.

Passioura, J. P. 1983. Roots and drought resistance. *Agric. Water Management* 7: 265-280.

Pugnaire, F.I., Serrano, L. & Pardos, J 1999. Constraints by water stress on plant growth. In: Hand book of plant and crop stress (Ed. M. Pessarkli), Marcel Dekker, Inc. New York. Basel. **PP:** 271-283.

Quarrie, S.A., Peki, S., Lazi-Jani, V., Ivanovi, M. & Vasi, G. 1995. ABA and drought toleranceerance in maize: future prospects. In: Maize Breeding, Production and Utilization: 50 years of Maize Research Institute, Zemun Polje (Ed. Z. Videnovic), Belgrade, Maize Research Institute. **PP:** 121-137.

Ramesh babu, V. & Rao, D.V.M. 1983. Water stress adaptations in the groundnut (*Arachis hypogaea* L.). Foliar characteristics and Adaptation to moisture stress. Plant Physiology and Biochemistry. 10(1): 64-80.

Ray, J.D., Yu, L.X., McCouch, S.R., Champoux, M.C., Wang, G. & Nguyen, H.T. 1996. Mapping quantitative trait loci associated with root penetration ability in rice (*Oryza sativa* L.). *Theoret. Appl. Genet.* 42**:** 627-636.

Robertson, B.M., Hall. A.E. & Foster, K.W. 1985. A field technique for screening genotypic differences in root growth. *Crop Science.* 25: 1084-1090.

Saxena, N.P., Johansen, C., Rheenen, H.A. Van, Saxena, M.C., Walby, B.J. & Hall, S.D. 1990. Chickpea ideotypes for genetic enhancement of yield stability in South Asia. In: Chickpea in the nineties: Proceedings of the second international workshop on chickpea improvement, 4-8 December 1989. (Eds. H.A. Van Rheenen, M.C. Saxena, B.J. Walby & S.D. Hall), ICRISAT, India. **PP:** 81-85.

Saxena, N.P., Krishnamurthy, L. & Johansen, C. 1994. Registration of a drought resistant chickpea germplasm. *Crop Science.* 33: 1424.

Schobert, B. & Tschesche, H. 1978. Unusual solution properties of proline and its interaction with proteins. *Biochem. Biophysactata*. 541: 270-277.

Serraj, R., Bidinger, F. R., Chauhan, Y. S., Seetharma, N., Nigam, S. N., Saxena, N. P., Kijne, J. W., Barker, R. & Molden, D. 2003. Management of drought in ICRISAT cereal and legume mandate crops. In: Water Productivity in Agriculture: Limits and Opportunities for Improvement (Eds. J.W. Kijne & R. Barker), CABI Publi., Wallingford, U.K. **PP:** 127-144.

Shackel, K.A. and Hall, A.E. 1983. Comparison of water relations and osmotic adjustment in sorghum and cowpea under field conditions. *Australian Journal of Plant Physiol.* 10: 423-435.

Sharma, K.K. & Lavanya, M. 2002. Recent developments in transgenics for abiotic stress in legumes of the semi-arid tropics. *Japan International Research Center for Agricultural Sciences No.* 23: 61-73.

Silim, S.N. & Saxena, M.C. 1993. Adaptation of spring-sown chickpea to the Mediterranean basin II. Factors influencing yield under drought. *Field Crops Research* 34(2) 137: 146.

Sinclair, T.R. & Ludlow, M.M. 1986. Influence of soil water supply on the plant water balance of four tropical grain legumes. *Australian Journal of Plant Physiol*. 13: 329-341.

Singh, K.B., Omar, M., Saxena, M.C. & Johansen, C. 1996. Registration of FLIP 87-59C, a drought resistant chickpea germplasm line. *Crop Science* 36: 472.

Singh, K.P. 1990. Evaluation of osmotic drought tolerance in chickpea (*Cicer arietinum* L.) varieties during germination and seedling growth. *Indian Agriculturist* 34(2): 103-106.

Singh, R. A. & Kumar, Prabhat 1992. Effect of water stress and hardening on metabolic changes in chickpea (*Cicer arietinum* L.). *Indian J. Plant Physiol.* XXXV (3): 252-257.

Srinivasan, A., Takedu, H. & Senboku, T. 1996. Heat tolerance in food legumes as evaluated by cell membrane thermostability and chlorophyll fluorescence techniques. *Euphytica* 88: 35-45.

Subbarao, G.V., Johansen, C., Slinkard, A.E., Rao, R.C.N., Saxena, N.P. & Chauhan, Y.S. 1995. Strategies for improving drought resistance in grain legumes. *Critical Review Plant Science* 14: 469-523.

Taiz, L. & Zeiger, E. 1998. Stress physiology. In: Plant physiology by L. Taiz and E. Zeiger. Sinauer Inc. Pub., Sunderland, Massachusates, **PP:** 725-757.

Turner, N.C. 1986. Crop water deficits: a decade of progress. *Adv. Agronomy* 39: 1-51.

Turner, N.C. 1991. Measurement and influence of environmental and plant factors on stomatal conductance in the field. *Agric. Forest Meteorol*. 54: 137-154.

Turner, N.C. & Begg, J.E. 1981. Plant Water relations and adaptation to stress. Plant Soil. 58: 97-131.

Turner, N.C. & Henson, I.E. 1989. Comparative water relations and gas exchange of wheat and lupins in the field. In: Structural and functional Responses to Environmental Stresses: Water shortage (Eds. K. H. Kreeb, H. Richter & T. M. Hinckley), The Hague, SPB Academic Publ. **PP:** 293-304.

Turner, N.C. & Jones, M.M. 1980. Turgor maintenance by osmotic adjustment: A review and evaluation. In: Adaptation of Plants to Water and High Temperature Stress (Eds. N.C. Turner and P. J. Kramer), New York, N.Y., John Wiley Publ. **PP:** 87-103.

Wery, J., Turc, O., Lecoeur, J., Singh, K. B. & Saxena, M. C. 1993. Mechanisms of resistance to cold, heat and drought in cool season legumes, with special reference to chickpea and pea. In: Breeding for stress tolerance in cool season food legumes (Ed. K. B. Singh), John Wiley & Sons Ltd., Chichester, U.K. **PP:** 271-291.

White, J. W. & Castillo, J. A. 1990. Studies at CIAT on mechanisms of drought tolerance in beans. In: Research on Drought Tolerancein Common Bean (Eds. J.W. White, G. Hoogenboom, F. Ibarra and S.P. Singh), Cali, *Centro International de Agriculture Tropical*. **PP:** 146-151.

Winicov, I. 2000. Alfin 1 transcription factor over expression enhances plant root growth under normal and saline conditions and improves salt tolerance in alfalfa. *Planta* 210(3): 416-422.

Wright, G. C., Hubick, K. T. & Farquhar, G. D. 1991. Physiological analysis of peanut cultivar response to timing and duration of drought stress. Australian *Journal of Agricultural Research*. 42: 453-470.

Zhang, J., Schurr, U. & Davies, W. J. 1987. Control of stomatal behavior by abscisic acid which apparently originates in the roots. *J. Exp. Bot.* 38: 1174-1181.

5

Production Technology in Chickpea

M.M. Singh, S.S. Shekhawat and Surendra Singh

For cultivation of any crop, production technology is very important because it is real cultivation of the crop. If we do not adopt appropriate method of cultivation we are not able to even germinate the crop. Therefore, for successful cultivation of the crop we have to take care the following points:

CLIMATE

Chickpea is cultivated during *rabi* in tropics and spring in temperate regions. In India it is adapted to cool dry condition and grown in post rainy season. It is tolerant to moisture stress and low management.

SOIL

Chickpea may be cultivated on a wide range of soils from sand dunes of Rajasthan to vertisols of Maharashtra. However, the loam soils with good water retention capacity and drainage are suitable for its cultivation.

FIELD PREPARATION

Soil should be friable, weed free, well drained and levelled, so that water does not stagnate in the field. In mono-cropping system when fields are kept fallow during monsoon deep ploughing before rains and 2-3 ploughing before withdrawal of rains helps in increasing profile moisture. If it is cultivated after *kharif* crop then one deep ploughing by soil turning plough followed by 2-3 harrowing and planking are sufficient. In case of less moisture pre-sowing irrigation should be given before field preparation, it will help in germination of the crop. If manure is to be applied, it should be applied before field preparation so that it may be mixed well in the soil.

CROP ROTATIONS

The important crop rotations are as under:-

1. Jowar/pearl-millet – chickpea
2. Maize – chickpea
3. Clusterbean – chickpea
4. Paddy – chickpea
5. Sesame – chickpea
6. Cotton/rice-chickpea
7. Groundnut - chickpea
8. Onion – chickpea

MIXED/INTER CROPPING

Chickpea is also cultivated as an inter crop or mixed with wheat, barley, linseed, safflower, coriander or mustard. In mixed/inter cropping, chickpea does well when rainfall is low. Intercropping is better than mixed cropping to perform agronomic practices well.

SEED TREATMENT

Seed should be treated with appropriate fungicide, insecticide and culture. It should be treated in FIR (fungicide then with insecticide (if any) and lastly with *Rhizobium*) manner. To overcome diseases like root rot and wilt seeds should be treated with Thiram 2.5 gm + Carbendazim @ 2 g or Thiram @ 3 g or Captan @ 2.5 g/kg of seed. To overcome the incidence of termite seed should be treated with Chlorpyriphos @ 10 ml/kg of seed. After treating with fungicide and insecticide seed should also be treated with *Rhizobium* culture, PSB and PGPR each @ 600 g/ ha seed at the time of sowing. *Rhizobium* culture enhances the nitrogen fixation through biological nitrogen fixation (BNF), PSB converts insoluble phosphates into soluble form & make it available to the plants and PGPR promotes the plant growth, which consequently increases the yield. It also improves seed quality. The *Rhizobium* culture, PSB and PGPR should be mixed well in *gur* solution and then mix with seed. Care should be taken that during mixing of culture, *gur* solution should not be much hot, because *Rhizobiums* are living organism and hot solution may kill the *Rhizobium*. After seed treatment the seed should be dried in shade for 6-8 hours and sowing should be done immediately or within 12 hours.

BIOLOGICAL NITROGEN FIXATION

Biological nitrogen fixation (BNF) supports sustainable and eco-friendly approach of nitrogen supply in legumes or it offers an economically attractive and ecologically sound means of reducing external inputs and improving internal resources. BNF is an important process in legumes which makes them more useful in the present energy crisis. The excessive use of chemical fertilizers often causes adverse affect on soil and environment and created so many problems like food safety, food quality, and human and animal health. Biofertilizer is a substance which contains efficient strains of microorganisms that colonize, when introduced in the rhizosphere or in the root zone of plants. The atmosphere contains about 78% of nitrogen which is primary source of nutrient for plants and other biological organisms. Bacteria of the genus *Rhizobium* play an important role in agriculture by inducing nitrogen-fixing nodules on the roots of legumes. *Rhizobium* is a type of nitrogen fixing bacteria that lives in soil and forms a symbiotic association with the root cells of leguminous plants or *Rhizobia* are soil bacteria that fix nitrogen after becoming established inside root nodules of legumes

(Fabaceae). This symbiotic relationship between the bacteria and the root tissue is mutualistic, menance both organisms benefit from the association. The *Rhizobium* appears to be host specific; in other words, one species of *Rhizobium* seems able to nodulate only one species of legume. It is common practice before planting the seeds of leguminous plants to inoculate with *Rhizobia* because not all soils contain the right species of *Rhizobium* for optimal symbiosis with a given legume.

Sulphur and phosphorous are nutrients known to improve both the quantity and quality of the produce. Pulse crops fix atmospheric nitrogen and require only small amount through fertilizers, however, they require as much as P and S even more than cereal crops. Phosphorous and sulphur application together appears to have synergistic or antagonistic effects. Furthermore, sustainable crop production needs judicious use of fertilizers along with bio-fertilizers. Conjunctive use of *Rhizobium* with phosphorous solubilizing biofertilizer (PSB) and plant growth promoting *rhizobacteria* (PGPR) revealed synergistic effect on symbiotic parameters and seed yield. PSB and PGPR improve the BNF by enhancing nodulation through colonizing root system and suppressing growth of deleterious microorganisms. *Rhizobium* and PSB known to enhance N and P content due to more nitrogen fixation by *Rhizobium* and solubilisation of native P by PSB. Further, it also makes them available to the plants by their synergistic effect. PGPR and *Rhizobium* causes growth stimulation of plant and ultimately enhances the yield. All the three inoculants *Rhizobium* + PSB + PGPR showed synergistic effect on symbiotic parameters and seed yield. Use of biofertilizer is able to increase the yield up to 20 per cent and can replace nitrogen and phosphorous fertilizers by 25 per cent.

WEED CONTROL

Being the short stature of the crop, chickpea suffers badly with weeds if not controlled in early stage. Weed management assumes greater significance because of seed standards and high value of seed. Clean cultivation with proper weed control during seed production makes subsequent cleaning and grading easier. Sometimes weeds act as a carrier for so many diseases and insect-pests, which deteriorate the quality and storability of seed. Seed production plot should be completely weed free because weeds adversely affect the production and quality of seed. Weeds compete with crop for moisture content, space, nutrients and sunlight and ultimately results in low yield. Weeds can be controlled mechanically and by chemical methods. Two hand weeding (at 25-30 DAS and 50-55 DAS) followed by 2-4 intercultural operations are quite effective in controlling the weeds. Pre-emergence application of Pendimethalin 30 EC @ 0.75-1 kg ai /ha. + one hand weeding at the time of 35-40 DAS and 2-4 intercultural operations are also equally effective in controlling weeds. Proper soil moisture should be ensured at time of application of weedicide in the field otherwise, it will adversely affect the crop.

IRRIGATION AND DRAINAGE

Though, chickpea is a rain-fed crop, but there must be proper arrangement for irrigation in seed production plot because in case of moisture stress or in drought if there is absence of irrigation facilities it will adversely affect the quantity and quality of the seed. Number of irrigation and their interval depends on whether condition and soil type. In case of insufficient moisture at the time of sowing, pre-sowing irrigation may be applied for proper germination. In case of failure of rains, irrigation should be applied time to time. Irrigation before flowering and at the time of pod filling improves the production significantly. Irrigation at the time of flowering is not desirable. Chickpea is sensitive to excess water so in case of heavy rains proper drainage should be there because in standing water, logging may occur and it may cause heavy loss to the seed production.

HARVESTING AND THRESHING

When plants become reddish brown, dry up and start shedding, harvesting should be done. Preferably harvesting should be done in the morning. The harvested plants are to be dried for a week for threshing. Threshing may be done manually or with the help of multi crop thresher with proper adjustment. Seed should be cleaned and dried in the sunlight so that seed may not damage in store.

YIELD

In rain-fed conditions the seed yield of chickpea is 8-10 q/ha, whereas, in irrigated conditions the yield levels reaches up to 25 q/ha.

REFERENCES

Ali, Masood and Mishra, J.P. 2004. *Rabi* pulses (Chickpea, lentil, lathyrus and frenchbean). Text book of field crops production. Directorate of information and publication of Agriculture by Rajendra Kumar. ICAR Krishi Anusandhan Bhawan, Pusa, New Delhi. **PP:** 317-371.

Ali, S.S and Gaur, H.S. 2007. Management of plant parasitic nematodes in pulses. Souvenir. National Symposium on Legumes for Ecological Sustainability: Emerging Challenges and Opportunities held at IIPR, Kanpur. Nov. 3-5, 2007 **PP:** 24-29.

Ali, S.S. and Singh, Bansa 2007. An overview of nematode constraints in pulse and their management. *Journal of Food Legumes*. 20(1): 1-8.

Bahl, P.N. 1988. Breeding approaches for self pollinated crops "Grain legumes". Plant Breeding by V.L. Chopra. Oxford and IBH Publishing Co. Pvt. Ltd. New Delhi, Bombay and Culcutta. **PP:** 121-137.

Chhabra, K.S.; Lal, S.; Kooner, B.S. and Verma, M. M. 1992. Insect pests of pulses. Identification and control manual. Punjab Agricultural University, Ludhiana and Directorate of Pulses Research, Kanpur.

Janoria, M.P.; Gour, V.K. and Singh, C.B. 1984. Perspectives in grain legumes. National Seminar on "New Dimensions in Pulse Research and Development" held at Jabalpur.

Khare, M.N.; Kumar, Kumud and Gurha, S.N. 2000. Management of Seed Borne Pathogens of Pulse Crops. Advances in Management of Biotic and Abiotic Stresses in Pulse Crops. Proceedings of National Symposium on "Management of Biotic and Abiotic Stresses in Pulse Crops" held at Kanpur, June 26-28, 1998. **PP:** 27-33.

Khedar, O.P.; Singh, R.V.; Shrimali, Mahesh and Singh, N.P. 2008. Pulses: Status and Cultivation Technology. Aaviskar Publishers, Distributors, Jaipur (Raj.)

Kooner, B.S.; Taggar, Gaurav K. and Cheeme Harpreet K. 2007. Insect Pests of Pulses-Status and Management. Pulses at a glance. Pulses Section, Department of Plant Breeding, Genetics and Biotechnology, PAU, Ludhiana-141004.

Mishra, S.P. 2004. Project Coordinator's Report. All India Coordinated Research Project on Chickpea. Indian Institute of Pulses Research, Kanpur. Annual Group meet held at JAU, Junagarh, Sept. 15-17, 2004.

Sekhon, H.S.; Singh, Guriqbal and Hari Ram 2007. Scope of pulses in crop diversification. Pulses at a glance. Pulses section, Department of Plant Breeding, Genetics and Biotechnology, PAU Ludhiana.

Singh, N.P.; Sewak, Shiv; Iquebal, M.A.; Chaturvedi, S.K. and Nath, Omkar 2009. Improved varieties of chickpea in India. AICRP on Chickpea, IIPR, Kanpur.

Singh, N.P.; Sewak, Shiv and Iquebal, Mir Asif 2009. All India Coordinated Research Project on Chickpea: A Profile IIPR, Kanpur.

Singh, S.J. 2001. Pulses production in Rajasthan. Souvenir, "Annual Group meet of AICRP on Chickpea and MULLaRP held at Agriculture Research Station, Durgapura, Jaipur, September 3-6, 2001. **PP:** 10-13.

Yadav, S.S. 2003. Strategies for increasing genetic yield potential of chickpea. Souvenir. National Symposium on "Pulses for Crop Diversification and Natural Resource Management" held at IIPR, Kanpur, 20-22 Dec. 2003 PP: 26-32.

6

Intregated Nutrient Management in Chickpea

Baldev Ram, S.S. Shekhawat, O.P. Khedar and A.S. Jat

India accounts for 2.2% of the global land and 16% of the world's population. Agriculture remains the main occupation in India and over 75% of its population generally depends directly or indirectly on agriculture for their livelihood. Every year about 19 million Indians are added to the existing population, which is currently more than 1 billion (approximately 1.18 billion). It is estimated that by 2010 about 246 million tonnes of food grains will be needed annually, and to produce this quantity, about 25 million tonnes of plant nutrient (N + P_2O_5 + K_2O) as mineral fertilizers would be required. The estimation for the year 2025, India will need at least 300 million tonnes (m t) of food grains, the production of which would require 45 m t of nutrients. It is estimated that required quantity only 35 m t will be met through chemical fertilizers and the remaining 10 m t will have to be met from organic manures, crop residues and Biofertilizers.

Indian agriculture presently faced food and nutritional security, environmental safety, energy crisis, global warming and climate changing scenario. The long term sustainability of agricultural productivity and environmental safety are being questioned due to yield stagnation, depleting soil organic carbon, low fertility status, nutrient mining, imbalance fertilizer use and infestation of pests and diseases. Soil degradation includes the loss of natural fertility for the production of food and regenerative raw materials. The most widespread phenomenon is the loss of organic matter and of the nutrients needed for biomass production as a result of improper soil management. Crop production under intensive cultivation over the years resulted in large scale removal of nutrients from the soil. Inadequate replenishment over years has resulted a negative nutrient balance and also called nutrient mining. Thus, excessive and imbalanced use of chemical fertilizers and pesticides not only deteriorated soil fertility but also poisoned the soil, water, micro flora, atmosphere, human beings and animals, causing several deformities, inabilities and serious diseases.

The consumption of cereals and pulses (*dal*) in the daily diet is considered nutritionally balanced. The cereals being poor in lysine is well complemented by the richness of the same in pulses. On the other hand, cereals are rich in methionine as compared to pulses. Besides being a rich source of protein, it maintain soil fertility through biological nitrogen fixation in soil and plays a vital role in furthering sustainable agriculture (Kannaiyan, 1999). Since time immemorial, pulses have been cultivated on marginal and sub-marginal lands, which are characterized by poor soil fertility and moisture stress, and consequently their yield potentials have not been realized. Further, more than 90 per cent area under pulses is under rain-fed. Therefore, there is a great scope of increasing the production in rain-fed areas through nutrient management. Nutrient imbalance is one of the major abiotic constraints of low productivity of pulses. The inbuilt mechanism of biological N_2 fixation enable pulse crops to meet 80–90 per cent of nitrogen requirements, hence a small dose of 15–25 kg N/ha is sufficient to meet out the requirement of most of the pulse crops. Phosphorus deficiency in soils is wide spread and most of the pulse crops have shown good response to 20–60 kg P_2O_5/ha depending upon nutrient status of soil, cropping system and moisture availability. Response to potassium application is location specific. In the recent years, use of sulphur (20–30 kg/ha) and some of the micronutrients such as Zn, B, Mo and Fe improved productivity of pulse crops considerably. Sulphur and phosphorous are nutrients known to improve both the quantity and quality of the produce. Sulphur provides a profound influence on protein synthesis. Thus, legumes have remarkable affinity for P and S. Phosphorous and sulphur application together appears to be synergistic or antagonistic effects. Furthermore, sustainable crop production needs judicious use of fertilizers along with biofertilizers.

The integrated nutrient management (INM) is the maintenance and possibly inhances soil fertility for increasing productivity through optimal use of all possible sources of plant nutrients of organic and inorganic which required for crop growth and quality in an integrated manner. Integrated nutrient management (INM) an eco-friendly approach includes use of organic manures, green manuring, incorporation of crops residues, inclusion of legumes in cropping system and biofertilizers along with the chemical fertilizers. Use of organic manure, green manuring, crop residues along with inorganic fertilizers not only reduces the demand of inorganic fertilizers but also increased the efficiency of applied nutrients due to its favourable effect on physical, chemical and biological properties of soil.

The main objective of the integrated approach to plant nutrient management is to coup-up all the major sources of plant nutrients in a judicious manner and to ensure their efficient use. The major sources of plant nutrients are (i) soil source (ii) fertilizer source (iii) organic source (iv) biological source. Among these four sources, the soil source needs careful manipulations so as not to deplete the plant nutrients by over exploitation through intensive cropping or through defective management. In fact, plant nutrients should be added to the soil through the other three sources in such a way that the nutrients removed by crops or cropping systems are less than that are added and there is a gradual increase in soil reserve. The inorganic fertili-zers should be used in such a way that there should be maximum use of the nutrient required by the plant with a minimum loss, since this is the most expensive input of the four sources. The organic source is the oldest source used by the farmers for supply of plant nutrients. However, this source has low nutrient content and has to be applied in bulk to meet the nutrient requirement of a crop. With the introduction of high nutrient requiring high yielding varieties, it has become virtually impossible to meet the complete nutrient requirement of a crop through these organic sources. The biological source includes green manuring and *in-situ* application of microbial inoculants etc. These sources can supply or make available a limited quantity of plant nutrients resulting into short fall of the requirements of high yielding varieties. In addition to the biological sources mentioned above, the adoption of a proper cropping sequence so as to conserve or slightly improve the nutrient reserve of soil can be accepted as another biological source for plant nutrient management. The integrated approach of nutrient management, thus aims a judicious use of all the four sources mentioned above in an integrated manner, taking into account the farming situation, ecological, social, economic factors of locality considering the ecofriendly approach.

CONCEPT OF INTEGRATED NUTRIENT MANAGEMENT (INM)

The basic concept of INM is the maintenance or adjustment of soil fertility and plant nutrients supply to an optimum level for sustaining the desired productivity through optimization of the benefits from all possible sources of plant nutrients in an integrated manner. Besides supplying all the nutrients in a holistic manner, organic manures such as compost, crop residues and FYM improves the soil organic matter (SOM), improving soil aggregation, water holding capacity, soil microbial population and thus intrances aggregation resulting into increasing water infiltration in the soil.

GOALS OF INTEGRATED NUTRIENT MANAGEMENT

1. To ensure productivity and sustainability of agriculture.
2. To maintain soil productivity and health.
3. To reduce the input cost by using farm manure and crop residues.
4. To utilize the potential benefits of green manures, leguminous crops and biofertilizers.
5. To prevent degradation of the environment.
6. To meet the social and economic aspirations of the farmers without harming the natural resource base of the agricultural production.
7. To save the ecosystem.
8. To coup-up with climate change and global warming.

PRINCIPLES OF INTEGRATED NUTRIENT MANAGEMENT

Integrated Nutrient Management (INM) refers to the practice of using optimum combination of different sources of nutrients supply (chemical fertilizers, organic manures, crop residues and biofertilizer's) for efficient crop production. There are five principles of INM for sustaining cropping system:

1. In the soil, replenishment of chemicals removed by the crop.
2. Maintenance of humus level in the soil, i.e. physical texture of the soil.
3. Avoidance of weeds, insect-pests and diseases.
4. Control of soil acidity and toxicity.

5. Control of soil erosion.

The role of soil biota in these five guiding principles of INM cannot be undermined, since soil microbes contribute in a big way to the soil organic matter dynamics, nutrient use and *in situ* control of pests and diseases.

OBJECTIVES OF INTEGRATED NUTRIENT MANAGEMENT

The naturally occurring biota in the soil may not be adequate for this purpose, so the judicious use of crop residues, microbial inoculants and chemical fertilizers in combinations may be needed to:

1. Enhance the availability of nutrients from all these sources in the soil for the growing season.
2. Meet out the requirement of the nutrients by the crop and act as a external sources.
3. Optimize various soil-biota interactions.
4. Minimize the loss of nutrients in the soil and also the hazardous effects of chemicals and fertilizers on the environment which helps in maintaining sustainability and ecosystem.

NEED FOR INTEGRATED NUTRIENT MANAGEMENT

The long term effects of chemicals and fertilizers on soil properties were observed during the 20th century and a need of integrated nutrient management (INM) was realized. The results of long term effects of chemical fertilizers on soil health attention to implement immediate use of INM due to: (i) Imbalanced use of N:P:K fertilizers have caused deleterious long term effects on soil fertility; (ii) In intensive cultivated areas, application of more chemical fertilizers (N:P:K) is not sufficient for sustaining the yields and it also leads to secondary and micro nutrients deficiencies in the soil which limits crop productivity; (iii) Use of organic manures, green manures, crop residues and biodegradable rural and urban waste not only supplement the chemical fertilizers, but also increases the efficiency in nutrients supply, leading to improvement in physical and biological properties of the soil. It is therefore, long term sustainable crop productivity can be achieved if optimum doses of N: P: K are supplemented with the application of 10 to 15 tonnes/ha, Farm Yard Manure annually.

COMPONENTS OF INTEGRATED NUTRIENT MANAGEMENT

Integrated Nutrient Management involves maintenance or adjustment of soil fertility and plant nutrients supply to an optimum level for the desired level of productivity. It can be achieved through optimization of benefits derived from different sources of nutrients supply.

A. Chemical Fertilizers

Out of 16 essential elements required for the nutrition of plants, pulses needs adequate amount of N, P, Ca, Mg, S, Mo and B. Pulse fixes atmospheric N to the soil. Recommended dose of nitrogen (N) and phosphorus (P) fertilizers are used but application of potassium (K) is generally neglected, resulting in to imbalanced nutrient supply leads lower yields. Nitrogen is required for the initial growth i.e. starter dose to stimulate the *rhizobial* activity as well as development of roots and root nodules. Phosphorus is required for proper growth of roots and *rhizobia*. Under intensive cropping systems, large amounts of K is removed; leading to serious depletion of soil K. Light textured alluvial soils, red and lateritic soils, and shallow black soils need immediate attention regarding K management for enhanced pulse crop productivity. Improved K supply is commonly associated with improved protein content in pulse grains, N fixation and water use efficiency. Calcium and magnesium are required to stimulate growth and to increase the size of the nodules, pod formation and seed setting whereas, Sulphur is required for nodulation and protein synthesis. Molybdenum is required for nitrogen fixation and assimilation whereas, boron helps in reproduction.

a. Nutrient Management

The results of field experiment on various pulses under cultivation rainfed showed that application of 20 kg K/ha along with nitrogen and phosphorus gives maximum response in the seed yield in comparison to higher dose of K, i.e. 30 and 40 kg/ha (Table 6.1) as reported by Tiwari and Tiwari (1999).

Table 6.1. Response of pulses for potash (K) on farmer's fields under rain-fed condition.

Crops	Seed yield response to K (kg/ha)			
	No. of trials conducted	**T1**	**T2**	**T3**
Chickpea	205	95	72	24
Urdbean	105	77	20	42
Lentil	90	112	85	73
Pigeonpea	69	163	29	59
Pea	15	148	87	81
Mungbean	14	30	29	-

T1= 20-40-20 ($N-P_2O_5-K_2O$ Kg/ha)
v/s 20-40-0 ($N-P_2O_5-K_2O$ Kg/ha)
T2= 30-60-30 ($N-P_2O_5-K_2O$ Kg/ha)
v/s 30-60-0 ($N-P_2O_5-K_2O$ Kg/ha)
T3=40-80-40 ($N-P_2O_5-K_2O$ Kg/ha)
v/s 40-80-0 ($N-P_2O_5-K_2O$ Kg/ha)

b. Foliar spray of nutrients

To mitigate the drought effects on the pulses spraying of cycocel at 100 ppm (100 mg/lit) is recommended. Cycocel spray enhances the root development, which facilitates the crop to get moisture from deeper layers. Dropping of flowers are commonly seen in pulses crops during drought situations. That can be check by application of NAA (4 ml/4.5 lit) spray is advocated, first spray at floral initiation followed by 15 days interval.

B. Organic manures

1. Farm yard manure (FYM)

This is the traditional manure and is mostly readily available to the farmers. It is rich in nutrients. A small portion of N is directly available to the plants while a larger portion is made available as and when FYM decomposes, when cattle dung and urine are mixed, a balanced nutrition is made available to the plants. Well rotten farm yard manure contains 0.4 to 1.5 per cent N, 0.3-0.9% P_2O_5 and 0.3-1.9% K_2O. Application of FYM improves soil fertility. It contains 0.5% nitrogen, 0.25% phosphorus and 0.4% potash. Manure contains stable and unstable forms of nitrogen. Unstable N occurs in urine as urea and may account for more than 50% of the total N in manure. Urea changes rapidly to ammonium ions then quickly to ammonia as pH increases and manure begins to dry. The more stable organic N occurs in the feces and is slowly released. Approximately 40-50% of the stable organic N will be available the first year, 12-15% of the N remaining the year after, 5-6% in the third year and lessor amounts in each subsequent year. Generally FYM is applied in pulses 10-15 t/ha to meet the nutrients requirement.

2. Vermicompost

Vermicompost is organic manure produced through bio-conversion of organic waste materials into nutritious compost by earthworm activity. Some specific earthworms act as bioreactors to decompose the wastes i.e. *Eisenia foetida*, *Eudrillus eugeniae* and *Perionyx excavatus* are the most suitable species used for this purpose. It is prepared in less time (generally 3-4 months) in comparison to the compost preparation by conventional methods. The compost produced includes vermicasts, as a source of available form of plant nutrients along with vitamins and growth promoting hormones. The nutrient contents on oven dry basis are N (1.2-1.8%), P (0.3%), K (0.6%) and organic carbon (18.2%). However, the quality will depend on the nature of the substrate and the stage of the decomposition. Vermicompost is used for the field crops @ 2.5 t/ha.

C. Crop residues

Most crops produce a voluminous amount of residue e.g. straw, stalk, husk etc. The estimated yield of crop residues is 185.30 m t per annum for important arable crops in India. The NPK content of the residue of different crops varies from 1.75 per cent to 3.30 per cent. The content of different nutrients in these residues varies depending on crop but it is estimated that they contribute about 1.24 m t of N, 0.16 m t of P and 2.00 m t of K. It is estimated that one third of the total quantity of residues is available for incorporation as residues in arable lands.

India has a vast potential of crop residues. Due to adoption of mechanical harvesting technology, a sizeable quantity of crop residues is left in the field which can form a part of nutrient supply. These crops are rice, wheat, soy-bean, mustard, sorghum, maize, pearlmillet, potato, sugar-cane and vegetables etc. Stubbles left in the field even in traditional harvesting methods ranged 0.5 to 1.5 t/ha in different crops if mechanical harvesting is done, this amount is much greater. There is a need to evolve appropriate man-agement practices to make use of these stubbles either by applying a part of chemical N recommended for the succee-ding crop during land preparation or by adding cellulose decomposing microbial cultures.

Mature crop residues being rich in lingo cellulose materials will help to maintain soil organic carbon (SOC) when incorporated into the soil. Decomposition of crop residues in the soil lowers the C/N ratio, which in turns increasing the mineralization of nutrients thus making them available to the plant. Crop residues requires to be chopped into smaller pieces for quicker absorption and microbial decomposition. The quantity of crop residues to be applied varies from 2.5-5.0 tonnes/ ha. These residues can be left on soil as mulch cum manure or incorporation of crop residues in soil may alter the physical and chemical properties of the soil and as a result can influence the yield. Leaving the crop residues in the soil may have positive, negative or no effect on seed yield depending on the condition of soil and climate. Counteracting these potential adverse effects of surface residues are beneficial effects like water conservation, soil erosion control, reduced winter killing and maintenance or enhancement of soil organic matter. Crop residues management practices also improves soil aggregation.

D. Enriched compost

The compost commonly prepared from the organic wastes is nutritionally imbalanced particularly in respect of the phosphorus. To make the compost balance, attempt has been made to apply single super phosphate (SSP) to the composting materials @25 kg SSP per tonnes of the compost. This practice not only enriches the product with P but also checks the volatilization loss of ammonical nitrogen. Phospho-compost can also be prepared out of paddy straw, sugarcane trash and other organic wastes. Generally paddy straw is chopped and soaked in water overnight for release of organic acid. Urea N @1% and Rock Phosphate @4% (Mussoorie, 100 meshes) is added to the straw material and inoculated with PSM (*Aspergillus awamori)* and the cellulose decomposing fungi (*Trichoderma viridae*) @ 1 kg per tonnes of compost. It is then covered with gunny bags; moisture is maintained at 60% level. After 60-75 days the material becomes ready for use. Enriched compost is used in pulse @ 1.25-2.5 t/ha as sole or in integrated manner.

E. Green manuring

Green manuring refers incorporation of a crop into the soil for its improvement. The advantages of green manuring for succeeding crops are economically feasible during the off-season or inter seeded with the main crop. It is made even more desirable when these crops include nitrogen fixing legumes.

1. In-situ green manuring

Dhaincha (*Sesbania cannabina*) and sunhemp (*Crotalaria juncea*) are mainly used for green manuring *in situ*. The crop is sown in the first fortnight of June by availing the pre monsoon rains. The seeds should be treated with 12 g of Sodium Molybadate and 1.2 g of Cobaltous Chloride 30 kg seed/ha. The entire quantity of phosphatic fertilizer of the succeeding crops is applied to the green manure crop. Green manuring results in vigorous growth, during their decomposition both N and P are available to the succeeding crop therefore, phosphate application is not needed. It is buried in the soil by ploughing after 6-7 weeks at the time of flower initiation. This practice adds about 12-15 t/ha green biomass which adds 60-70 kg /ha N. Cowpea and cluster bean can also be used as green manuring crops. Hollow stem of the rostrata species creates problem because it floats when chopped to pieces while irrigation. Supplementation of green matter and the N content (on dry weight basis) of green manuring crops are given in Table 6.2.

Table 6.2. *In-situ* green manure crops biomass addition and their N content on dry weight basis

Green manure crops	Green matter (t/ha)	Moisture (%)	N content (% of dry wt)
Crotolerancearia juncea (Sunhemp)	16.50	73.00	3.01
Sesbania cannabina (Dhaincha)	14.80	78.00	2.67
Vigna unguiculata	10.00	85.00	2.63
Cyamopsis tetragonoloba (Guar)	7.00	60.00	3.53

2. Ex-situ green leaf manuring

The green vegetative parts of the trees or bushes are collected and incorporated in the soil at the time of field preparation @ 5-6 t/ha. Plant species like *Pongamia pinnata, Pongamia glabra, Gliricidia maculata, Cassia tora, Sesbania speciosa and Ipomoea cornea* can be selected for green leaf manuring. Weeds like *Croton sparsiflorus, Leucas aspera* and Thornless *mimosa* are also utilized for this purpose.

Incorportaion of green leaf biomass and N content on (dry weight basis) of green manuring crops are summarized in Table 6.3.

Table 6.3. Ex-*situ* green leaf manure crops biomass addition and their N content on dry weight basis.

Green leaf manure crops	Green matter (t/ha)	Moisture (%)	N content (% of dry wt)
Gliricidia maculate	3.00	75.00	3.46
Sesbania punctata	3.70	73.00	2.42
Cassia tora	5.20	71.00	2.13

F. Nitrogen fixing crops

A legume contributes a substantial amount of nitrogen in cropping system. Generally, legumes fixes atmospheric nitrogen 30-60 kg N/ha in grain legumes and 90-120 kg N/ha in forage legumes. These are endowed with many desirable characters i.e. short duration, restorative (soil fertility building crops), low water requirement and suitable to be grown in mixed/intercropping systems and also as a catch crop to scavenge the residual soil moisture and fertility. The legumes improve soil fertility through nitrogen fixation. The nitrogen fixing values estimated for various legumes (Peoples *et al*. 1995) are given in Table 6.4.

G. Bio-fertilizers

The root system is always in close association with microorganisms and other sources of nutrients. The microbes in root zone are maintained due to secretions from the roots and are known as 'rhizosphere'. These microbes supply nutrients to the soil through their heterotrophic effect and supply nitrogen, phosphorus. Crop productivity and nutrient cycles are integral parts of the exploitation of soil health which causes soil degradation through nutrient depletion and erosion, so long term strategies are needed to curtail the use of chemical fertilizers without adversely affecting crop productivity.

Table 6.4. Estimates of nitrogen fixed by chickpea (Peoples *et al.* 1995)

Crop	Nitrogen fixed (Kg /ha)
Blackgram	119.00-140.00
Chickpea	23.00-97.00
Cluster bean	37.00-196.00
Cowpea	9.00-125.00
Greengram	50.00-66.00
Pigeonpea	4.00-200.00
Soybean	49.00-450.00
Peas	46.00

1. Rhizobium

Rhizobium is gram negative rod shaped bacteria, which invade the roots of legume, forming nodules and fixing the atmospheric nitrogen through nitrogenase enzyme which is the basic phenomenon of the biological nitrogen fixation. In leguminous plants on symbiotic association with soil bacterium, Rhizobium forms nitrogen fixing root nodules which are agronomically significant as it provides an alternative to the use of energy expensive nitrogenous chemical fertilizers. The symbiotic association between legume and *Rhizobium* may reduce about 70 million tonnes of atmospheric nitrogen to ammonia per annum which accounts to 40% of all biologically fixed nitrogen per year (Burns and Hardy, 1975).

The cross inoculant strains root nodulating rhizobia has been modified since 1984 from a wide variety of leguminous plants. A new system of classification was proposed by Jordan (1984) in Bergy's Manual of Systematic Bacteriology and in given is Table 6.5.

Table 6.5. Classification of root nodule bacteria (Jordan, 1984).

Genus	Species	Biovars	Host legumes
Rhizobium	*R. leguminosarum*	*viceae*	Vicia
R.leguminosarum	*trifolii*	-	Trifolium
R.leguminosarum	*phaseoli*	-	Phaseolus
R. meliloti	-	-	Medicago
R. loti	-	-	Lotus
Bradyrhizobium	*B. japonicum*	-	Glycine
Bradyrhizobium sp.	-	-	Gicer, cajanus, vigna

Rhizobium inoculation

Biological nitrogen fixation by symbiotic associations of such plants with micro organism is more economic and eco-friendly than nitrogenous fertilizers. The extensive survey of legume root nodulation showed that lack of adequate nodulation of grain legumes by an indigenous population of *Rhizobia*. The *Rhizobium* population is drastically reduced in the soil as most of the winter pulses follow kharif rice which is grown under anaerobic condition. It is needed to inoculate seeds with *Rhizobium* bio-fertilizer. Symbiotic nitrogen fixation by *Rhizobium* with nodulated legumes contributes significantly to the total biological nitrogen fixation. On an average 30 kg/ha nitrogen can be fixed by use of such inoculants in grain legumes, although the amount is much higher under favourable conditions. It is a cheaper method to meet out the nitrogen demand of the crop. The yield of the grain legume can be increased significantly by inoculating with right strain of *Rhizobium*. Approximately, 20 g/Kg *Rhizobium* culture is required to inoculate seeds. Bengal gram requires 1.0 kg/ha of culture. Pulse seeds treated with specific strains of Rhizobium increases (favourable CN ratio is 10:1) the seed yield through better nodulation and maintenance of organic matter in the soil (Saxena and Tilak, 1999).

Crop response to *Rhizobium*

The application of *Rhizobium* as seed inoculant is the essential component in the integrated nutrient management in pulses. The results showed that at least 25 per cent of nitrogen could be complemented by *Rhizobium* if included as a component in the integrated nutrient management.

2. Phosphate solubilizing bacteria (PSB)

Phosphorus (key element) is an essential plant nutrient required for early establishment and better plant growth. It accelerates tillering, flower initiation, good pod and seed setting. Majority of the Indian soils are deficient in available form of P and its requirement is meet out by the addition of phosphatic fertilizers but the efficiency of applied phosphorus rarely exceeds 30 per cent due to its fixation as Fe and Al phosphates in acid soil whereas, Ca and Mg phosphates in alkaline soils. Phosphate solubilizing bacteria play a major role in the solubilization and uptake of native as well as applied soil phosphorus. The bacteria, *Bacillus megatrium, B. polymyxa, Pseudomonas striata, Micrococcus, Streptomyces* are the important phosphobacteria commonly present in the soil. The population of these bacteria is more in the rhizosphere compared to non rhizosphere. These microorganisms produce organic acid which is accompanied by the acidification of medium to solubilize insoluble phosphorus compounds. Apart from acid production, it produces phosphatase enzyme which causes the solubalization of P.

Interaction of *Rhizobium* and Phosphobacteria

Combined inoculation of Rhizobium with phosphobacteria (*Bacillus megaterium* and *Pseudomonas striata*) for red gram, black gram, green gram and Bengal gram are used for increasing seed yield. Maximum seed yield was reported by the combination of rhizobial strain with phosphobacteria with full dose of N and P in red gram (986 kg/ha). It increased 34.70 per cent higher yield than the uninoculated control. Dual inoculation with half dose of fertilizer recorded seed yield of 880 kg/ha i.e. 20.22 per cent higher than control. Trials conducted at Coimbatore reported that combined inoculation of *Rhizobium* and phosphobacteria with 50 per cent of N and P fertilizer recorded the equal yield of 100% N and P alone. The results clearly indicated that combined inoculation of *Rhizobium* and Phosphobacteria helps the macro symbiont pulses to get maximum quantity of N and P nutrients which enhances seed yield.

Role of nutrients in BNF

Application of organic manures in pulses is essential for improving the nodulation and seed yield by *Rhizobium* in acid soils reported by Prabakaran (1998). Experimental findings depicts that soil application and foliar spray of P, Mo, Co, B and Fe recorded the maximum nodulation and enhances seed yield of urdbean.

CONSTRAINTS IN ADOPTION OF INM TECHNOLOGY

1. Non-availability of FYM.
2. Difficulties in growing green manuring crops.
3. Non-availability of qualitative biofertilizers.
4. Non-availability of soil testing facilities.
5. High cost of chemical fertilizers.
6. Non-availability of water.
7. Lack of knowledge and poor advisory services.
8. Soil/environmental conditions.
9. Non-availability of credit facilities.
10. Lack of awareness for INM technology.

CONCLUSION

Integrated use of FYM at 2.5 t/ha and 50% recommended NPK fertilizers plus *Rhizobium* and phosphobacteria inoculation saves 50% chemical fertilizers. It is true that pulses productivity can increased and sustained through the integrated approach of nutrient management along with the introduction of high potential pests and diseases resistant genotypes, proper seed rate, executing optimum time of sowing, and practicing IWM techniques. Coexistence of exhaustive pulse research, giving primary importance to chickpea and extension services will pave the path to meet out the demand of pulses for increasing population.

REFERENCES

Burns, R.C. and Hardy, R.W.F. 1975. Nitrogen fixation in bacteria and higher plants. Springer Verlag, New York **PP:** 189.

Gupta, J.P. and Gupta, G.N. 1986. Effect of tillage and mulching on soil environment and cowpea seedling growth under arid conditions. *Soil Tillage Res*. 7: 233-240.

Hooker, M.L.; Herron, G.M. and Penas, P. 1982. Effect of residue burning, removal and incorporation on irrigated cereal crop yields and soil chemical properties. *Soil Sci. Am. J*. 46: 122-146.

Illmer, P. and Schinner, F. 1992. Solubilization of inorganic phosphate by microorganisms isolated from forestry soils. *Soil Biol. Biochem*. 24: 389-395.

Jordan, D.C. 1984. Rhizobiaceae. **In**: *Bergey's Manual of Systematic Bacteriology* (eds. N. R. Kreig and J. G. Holt) Vol. I. Williams and Wilkins Publications, Baltimore.

Kannaiyan, S. 1999. Bioresources Technology for Sustainable Agriculture, Associated Publishing Co., New Delhi. **PP:** 422.

Prabakaran, J. 1998. Effect of pressmud and sheepmanure on the growth and yield of redgram in alfisols. *Madras Agric. J*. 85: 303-304.

Prabakaran, J.; Balachandar, D. and Nagarajan, P. 1999. Influence of dual inoculation of rhizobium and phosphobacetria at different levels of P in horsegram. *Legume Res*. 22: 183.

Peoples, M.B., Ladha, J.K. and Heraidge, D.F. 1995. Enhancing legume nitrogen fixation through plant and soil management. *Plant Soil*. 174.

Ramamoorthy, K. and Arokia, A.R. 1997. Agronomic effectiveness of organic sources and MRP to phosphorus economy in rainfed greengram. *Madras Agric. J*. 84(10): 593-595.

Ramamoorthy, K., Balasubramanian, A. and Arokia, A.R. 1997. Response of rainfed blackgram (*Phaseolus mungo*) to phosphorus and sulphur nutrition in red lateritic soils. *Indian Journal of Agronomy*. 42(1): 191-193.

Rao, U.M.B. and Singh, S.P. 1993. Crop response to organic sources of nutrients in dry land conditions. **In:** *Recent Advances in Dryland Agriculture* (ed) L.L. Somani, Scientific Publishers, Jodhpur. **PP:** 287-304.

Saxena, A.K. and Tilak, K.V.B.R. 1999. Potentials and prospects of rhizobium biofertilizer. **In**: *Agromicrobes* (eds M.N. Jha, S. Sriram, G.S. Venkataraman and S.G. Sharma), Todays and Tomorrows Printers and Publishers, New Delhi **PP:** 51-78

Tiwari, V.N. and Tiwari, K.N. 1999. Proceedings PPIC-UPCAR Workshop on use of Potassium in Uttar Pradesh Agriculture (Eds. Tiwari, K. N. and Modgel, S. C.). **PP:** 79-87.

7

Role of Bio-Fertilizers in Chickpea

Jagdish Prasad, S.S. Shekhawat, Kuldeep Sharma, O.P. Khedar and B.S. Shekhawat

Biological nitrogen fixation (BNF) supports sustainable and eco-friendly approach of nitrogen supply in legumes white offering an economically attractive and ecologically sound means of reducing external inputs with improvement in internal resources. BNF is an important process in legumes which makes them more useful in the present energy crisis. The misuse or excessive use of chemical fertilizers have often adverse affect on soil and environment which creates many problems like food safety, food quality health of animals and human beings. Bio-fertilizer is a substance which contains efficient strains of microorganisms that colonize, when introduced in the rhizosphere or in the zone that surrounds the root of plants. Inoculation of legume with beneficial bacteria started centuries ago and farmers experienced that when soil from the previous legume field when mixed with the fields where legumes are to be grown the yield increases. By the end of 19th century the mixing of naturally inoculated soil with seeds become a recommended practice in USA. The atmosphere contains about 78% of nitrogen which is primary source of nutrient for plants and other biological organisms. Bacteria of the genus *Rhizobium* play an important role in agriculture by inducing nitrogen-fixing nodules on the roots of legumes. *Rhizobia* are soil bacteria that fix nitrogen after becoming established inside root nodules of legumes (Fabaceae). This symbiotic relationship between the bacteria and the root tissue is mutulistic, means both organisms benefit from the association. The plant root cells provide nutrients and carbohydrates for energy to the bacteria and the bacteria reciprocate by fixed nitrogen compounds to the plant. This area near the roots is called the "*rhizosphere*" which is the thin layer of soil that sticks to the roots. The *rhizosphere* is a huge habitat in the soil, because plants have so many root fibers. It is common practice before planting the seeds of leguminous plants to inoculate with *Rhizobia* because not all soils contain the right species of *Rhizobium* for optimal symbiosis with a given legume. *Rhizobium* provides the major biological source of fixed nitrogen in agricultural soils. *Rhizobia* convert atmospheric nitrogen into nitrogen compound that the plants can use, this conversion of nitrogen is called "nitrogen fixation" and the *Rhizobia* live in special root tissues called "nodules".

Sulphur has a profound influence on protein synthesis. Thus, legumes have remarkable affinity for P and S. Pulse crops fix atmospheric nitrogen and require only small amount through fertilizers, however, they require as much as P and S even more than cereal crops. Phosphorous and sulphur application together appears to have synergistic or antagonistic effects. Furthermore, sustainable crop production needs judicious use of fertilizers along with bio-fertilizers. Conjunctive use of *Rhizobium* with phosphorous solubilizing bio-fertilizer (PSB) and plant growth promoting *rhizobacteria* (PGPR) revealed synergistic effect on symbiotic parameters and grain yield. PSB and PGPR improve the BNF by enhancing nodulation through colonizing root system and suppressing growth of deleterious microorganisms. *Rhizobium* enhance N through nitrogen fixation and PSB convert insoluble P in to soluble form. They also make them available to the plants by their synergistic effect. PGPR and *Rhizobium* causes growth stimulation of plant and ultimately enhances the yield. All the three inoculants *Rhizobium* + PSB + PGPR showed synergistic effect on symbiotic parameters and grain yield. *Bacillus* spp. and *Pseudomonas* spp. are efficient PGPRs. Though, the production and application of biofertilizers for leguminous crops are very common in India, but cost effective, ecofriendly and suitable technology is necessary to replace chemical fertilizers. Use of biofertilizer is able to increase the yield up to 20 per cent and can replace nitrogen and phosphorous fertilizers by 25 per cent.

FACTORS AFFECTING NODULATION AND N_2 FIXATION

A. Seasonal variation

N_2 fixation is more up to pod filling and decreases at maturity.

B. Environmental factors

Nitrogen fixation activity is maximum in the morning hours and decreases during dark periods.

C. Agronomic practices

Intercropping with cereals and deep sowing results in poor nitrogen fixation.

D. Nutritional factors

Increase in nitrogen fertilizer application decreases nitrogen fixation.

E. Application of plant protection chemicals

Some of the chemicals do not have any effect on nodulation such as Carbofuran, whereas, some of the chemicals such as Thimet reduces nodulation.

F. Crop species and variety

Crop species and variety play an important role in nodulation and nitrogen fixation. Some of the varieties are more nodulating and other are poor one.

BENEFITS OF USE OF *RHIZOBIUM*

Nitrogen fixation depends on many factors like crop, soil, crop growth and status of *Rhizobium* present in the soil. In general chickpea fixes atmospheric nitrogen up to 120 kg/ha/year.

On the basis of nature and function, biofertilizers are grouped in four classes.

A. Nitrogen fixing biofertilizers or symbiotic *Rhizobium*

Rhizobia are gram negative, rod shaped bacteria grouped under α-proteobacteria. They possess hydroxyl butyrate granules as stored food. They are motile when young. They produce acid or alkali on YEMA medium. They survive in soil and rhizosphere of legume as well as non legumes. Rhizobium secretes extra cellular polysaccharides which help in binding soil particles. They can survive at low temperature and also toleranceerate temperature up to 50°C.

Selection of rhizobial strains for inoculants production

A large scale screening should be carried out to identify ideal strain for specific crop. Though, the root-nodule forming bacteria are present in most of the soils but they vary in number and effectiveness in nodulation and nitrogen fixation. They are symbiotic nitrogen fixers that can be found in the roots of plants.

B. Phosphorous solubilizing biofertilizer (PSB)

Next to nitrogen, phosphorous is vital nutrient for plants and microorganisms. Several bacteria like: *Bacillus megaterium var. phosphaticum, Bacillus subtilis, B. circulance, Pseudomanas striata and Fungi: Aspergillus awamori, Penicillum* sp. possess the ability to convert insoluble phosphates in soluble form by secreting acids such as formic, acetic, propionic and succinic acids. PSB solubilise inorganic insoluble phosphorous by excreting organic acids that directly dissolve phosphatic materials

C. Phosphorous mobilizing biofertilizers

Arbuscular mycorrhiza-Acaulospora sp., *Glomous* sp.

D. Plant growth promoting Rhizobacteria

The bacteria that colonize roots or rhizosphere soil and are beneficial to crops are referred as PGPR such as Pseudomonads - *Pseudomonas flurescence, Pseudomonas* sp., *Bacillus subtilis.* The PGPR are referred as biostimulants produces the phytohormones such as indoleacitic acid, cytokinins, gibberellins and inhibitors of ethylene production. They also produce siderophores which chelate iron and pathogens are deprived for want of iron nutrition.

PRODUCTION OF BIOFERTILIZERS

Commercial production of rhizobial inoculants in India started at Indian Agricultural Research Institute, New Delhi in 1956. It was followed by State Agricultural Universities. Later state governments started production of biofertilizers to full fill the requirement of legume growers, but till date we are lacking in good quality biofertilizers in sufficient quantity. In 1978 Indian Council of Agricultural Research started All India Coordinated Research Project on Biological Nitrogen Fixation (AICRP-BNF). In 1983 National Project on the Development and use of Biofertilizers was launched by government of India. In 1988-89, Department of Biotechnology, Government of India initiated a massive project on Technological Development and Demonstration of Biofertilizers. Presently more than 95 firms (public and private) are engaged in the production of biofertilizers with an annual capacity of 18,000 metric tons. The use of quality and viability of *Rhizobium* inoculants is of prime importance to harness the desired yield of legumes. Though, the demand of inoculants is increasing day by day, but still does not meet the demand. Moreover, these are not readily available as the chemical fertilizers. National Biofertilizer Development Centre (NBDC), Agriculture Universities, State Government and Private companies are some of the agencies engaged in manufacture and promotion of biofertilizers in the country. For the production of inoculants following points should be take care:

A. Selection of *Rhizobial* strain

Different *Rhizobium* spp. is responsible for nitrogen fixation in different crops, so according to crop the *Rhizobial* strains should be used.

B. Preparation of inoculants

The inoculants may be prepared as per specification in controlled condition otherwise it is of no use, because *Rhizobiums* are living organism and they remain alive in specific condition only. The method of preparation is as under:

Medium

Yeast Extract Mannitolerance broth (ingradients g/l)

Yeast Extract	:	1.00
Mannitolerance	:	10.00
Sodium chloride	:	0.10
Calcium carbonate	:	1.00
Dipotassium sulphate	:	0.50
Magnesium sulphate	:	0.20

a. Aeration

Rhizobiums are living organism, so aeration is must for biological activities. In unaerobic condition they will not survive and such culture is of no use.

b. Inoculation level

The inoculation level is also dependent on the *Rhizobium* spp.

C. Selection of carrier material

The inert material used for carrier should be with good water holding capacity. It should support the viability of the *Rhizobium* for a long time.

QUALITY CONTROL

Quality control of *rhizobial* inoculants is an important aspect for ensuring the desired strain of *rhizobia* in sufficient numbers for effectiveness and survival in carrier, soil or seed. Despite existence protocols, only 20% inoculants sold in India are of exceptable quality. The Bureau of Indian Standard (BIS) has prescribed following ISI standards for *Rhizobium* inoculants.

1. Carrier should be inert with good water holding capacity.
2. *Rhizobium* inoculants should contain at least 10^8 viable cells/g of carrier.
3. Packing should be economically durable and easy to handle.
4. Precautions regarding storage and date of expiry should be mentioned on the packet.
5. Inoculants packet for a particular crop should be mentioned clearly.

STORAGE OF INOCULANTS

The storage of inoculants depends on carriers and strains of *Rhizobium*. In general survival of *rhizobia* is better if stored a 4-5°C. Storage in clay pots in shade is also effective for better survival of *Rhizobia*. Charcoal carrier based inoculants may be stored up to three months in the refrigerator. Shorter shelf-life, contamination in carrier and non availability of protective carrier material are some of the reasons responsible for poor quality of *Rhizobium* inoculants.

METHODS OF *RHIZOBIAL* INOCULATION

A. Seed pelleting

The usual method of pelleting involves addition of 40% of gum arabic or 5% carboxy methyl cellulose to inoculants slurry before application to seed. Finally ground calcium carbonate capable of passing through 300 mesh sieve is added to freshly inoculated wet seed in a container and mixed rapidly for 2 minutes until seeds are evenly coated. The lime coated seeds are white and are allowed to harden for an hour by spreading of surface. The seed may be sown immediately but it can be stored up to 2 weeks at low temperature (4°C).

B. Seed treatment

For all legumes culture is applied for seed inoculation. Carrier based cultures are mixed with gur or gum to form a slurry and seeds are added to the slurry so as to have uniform coating of culture. The seed should be treated with *Rhizobium* culture, PSB and PGPR@ 600 g/ ha at the time of sowing. After seed treatment with *Rhizobium,* PSB and PGPR the seed should be dried in shade for 6-8 hours and sown immediately or within 12 hours.

Table 7.1. Quantity of gur and water for the preparation of gur solution.

S. No.	Water (ml)	Gur (g)	Crop
1.	1000	250	Mungbean, urdbean cowpea, moth
2.	1500	300	Pigeonpea
3.	2500	300	Groundnut

PRECAUTIONS TO USE INOCULANT

1. Appropriate quantity of culture should be used.
2. Use of binding material is essential.
3. Store the inoculants (culture packets) in the cool place.
4. Dry the inoculated seeds out of direct sunlight in a cool shaded place.
5. Sowing should be done, preferably on the same day of seed Inoculation with *rhizobia*.
6. If the sowing is done on different days, the whole packet of the inoculants should not be used at once. Remove the required quantity of the inoculants, seal the packet with sticky tape and store in a cool place (preferably at 4-5°C).
7. For reliable results, the inoculants should not be used after the expiry date.
8. Slurry inoculated seeds should never be mixed with acidic fertilizers as it will kill the bacteria.
9. Good quality inoculants should contain at least 10^8 *rhizobia* per gram of carrier.

REFERENCES

Ahlawat, I.P.S. and Srivastava, T.K. (1997). Fertility management in pulse based cropping systems. Recent advances in pulses research (Asthana, A.N. and Ali, M.). Indian Society of Pulses Research and Development, IIPR, Kanpur. **PP:** 509-524.

Ali, M. 1989. Consolidated Report on Kharif Pulses. Directorate of Pulses Research, Kanpur.

Dudeja, S.S and Dulhan, J.S. 2005. Biological nitrogen fixation research in pulses with special reference to mungbean and urdbean. *Indian Journal of pulses research.* 18(2) 107-118.

Gopalaswami, G. 2012. Effective use of biofertilizers for sustained pulse production. Pulses production Strategies. by Dr. P. Murugesa Bhoopati and Dr. M. Paramathma. Sri Shakthi Promotional Litho Process. S.F. No. 283 Masaniamman

Nagar, Anna Nagar East, Edarpalayam, Coimbatore- 641025. **PP:** 457-476.

Khedar, O.P.; Maloo, S.R. and Punia, S.S. 2012. Pulses in Rajasthan: An overview Pulses production Strategies. by Dr. P. Murugesa Bhoopati and Dr. M. Paramathma. Sri Shakthi Promotional Litho Process. S.F. No. 283 Masaniamman Nagar, Anna Nagar East, Edarpalayam, Coimbatore- 641025. **PP:** 191-226.

Khedar, O.P.; Singh, R.V.; Shrimali, Mahesh and Singh, N.P. 2008. Pulses: Status and Cultivation Technology. Aaviskar Publishers, Distributors, Jaipur (Raj.)

Pal, K.K.; Dey, R. and Kulkarni, J.H. 2004. Biofertilizers for enhancing croundnut production. Groundnut research in India (edited by M.S. Basu and N.B. Singh) National Research Centre for Groundnut (ICAR) PB %, Junagarh 362001, Gujarat, India. **PP:** 339-347.

Pareek, R.P. and Chandra, R. 2005. Biological nitrogen fixation. Pulses (Singh, Guriqbal, Shekhon, H.S. and Kolar, J.S.) Agrotech Publishing Academy, Udaipur. pp 279-312.

Sharma, Poonam; Khanna, Veena and Shekhon, H.S. 2007. Biological Nitrogen Fixation: An Asset in Pulses. Pulses at a glance. Pulses Section, Department of Plant Breeding, Genetics and Biotechnology, PAU, Ludhiana-141004.

Singh, Pradeep Kumar; Ghos, Jyotirmaya; Kumar, Yogesh; Barnwal, Manoj Kumar; Thakur, R.K. and Karmkar, S. Improved cultivation of Pigeonpea in Jharkhand. Birsa Agricultural University, Technology Bulletin 28.

8

Integrated Weed Management in Chickpea

Baldev Ram, S.S. Shekhawat, O.P. Khedar, G.L. Yadav and V.D. Fageria

Weed management is as old as agriculture itself. Only concept and methods have changed over the years. Herbicides since the discovery of 2,4-D have revolutionalized agriculture and have contributed enormously in increasing food production. Weeds and crop plant share identical needs for natural resources. Weeds cause great loss to the crop. The weeds have the mechanism of wider adaptability in the adverse condition and high growth rate as compared to cultivated crops. Chickpea (*Cicer arietinum* L.) commonly known as gram or Bengal gram is most important pulse crop of the country. Weed species have different effects on yield losses of chickpea depending on their growth habit, nutrient requirement and water uptake. Chickpea crop faces competition mainly from annual broad leaf weeds due to identical growth pattern of chickpea and weeds. Being the short stature chickpea suffers badly with weeds if not controlled in early stage. Sometimes weeds act as a carrier for so many diseases and insect-pests, which deteriorate the quality and storability of produce. Weeds compete with crop primarily for moisture content, space, nutrients and sunlight and ultimately results in low yield. However, the loss in yield depends on various factors such as type and quantum of weed flora, duration of crop weed competition, variety etc. For good production, crop should be weed free. Innovative approaches to manage the weeds are in great demand around the world, particularly those which are less expensive and harmful to environment. The search for new control methods, which are effective and economic and have minimal side effect is a continuous process.

CONCEPT OF WEED

Weeds are unwanted and undesirable plants which interfere with the utilization of land and water resource and thus adversely affect human welfare. They can also be referred as plants out of place. The first person to use the word weed in the present day meaning was Jethro Tull, who is popularly known as father of tillage. The critical period of weed competition in pulse crops is generally during the first 30 days after sowing.

PREVENTIVE WEED MANAGEMENT FOR WEED FREE CROP

To obtain weed free crop following cultural and mechanical measures are to be adopted:

a. Proper cleaning of produced crop seeds.

b. Separating crop seeds from ad mixture of weed and crop seeds on the basis of physical difference like seed size, weight, shape and colour.

c. Weed seeds adhere to harrows, ploughs, intercultural equipments, hoes during tillage and post planting operations and get carried to other fields. Hence, agricultural equipment should be cleaned before using them.

d. During combine harvesting weeds seeds are also harvested along with crop which should be cleaned properly.

e. Many crop seeds can pick up weed seeds during uncovered transport so before transportation seed should be covers properly.

f. In India most of the peasants use their own saved seeds, which are full of weeds. Such seed should be clean properly.

g. Farmers have traditional processing methods, which cannot remove the weed seeds that are equal in size of crop seed. In such cases weeds creates problem so proper processing is required for weed free seed.

h. Grain/seed rejects containing weed seeds either reach the soil directly or fed to animals. The weed seeds consumed by cattle while grazing pass through the digestive canal without loss of their viability. These seeds germinate from the droppings and spread in other areas. Such wastes should be decomposed well in the FYM or compost pit rather than putting them directly in the fields.

i. Compost and FYM are the notorious source of weed seeds. Most of the weed seeds remain viable in the manure pits and spread like wild fire wherever the manures are applied. Using well-decomposed weed free farmyard manure helps to overcome the risk of spreading the weeds seeds.

CRITICAL PERIOD OF WEED CROP COMPETITION AND WEEDING

The critical period of weed crop competition is between early growth during which weeds can grow without affecting crop yield and the point after which weed growth does not affect yield. In general initial growth of chickpea is slow means there is strong competition in early stages, whereas, in later stages legumes grow faster means there is less competition with weeds. Timely weeding is an important aspect for weed management and seed yield too. Unfortunately, most farmers assume that weeding at any time during the growing season is enough for obtaining good crop yields. But it is a fact that time of sowing is as important as weeding itself. Weeds must be controlled with in the critical period of weed crop competition. Intercropping also reduces the weed competition as compared to sole crop.

LOSSES AND EXTENT

Weeds compete with crop primarily for moisture, nutrients, sunlight and space. It is widely known that losses caused by weeds exceed the losses from any other category of agricultural pests like insects, nematodes, diseases, rodents etc. Actually the weeds cause damage in terms of yield losses, uptake of nutrients and deterioration of seed quality. Of the total annual loss of agricultural produce from various pests in India, weeds account for 45% insect 39%, disease 30% and other pests 5%.

Table 8.1. Yield losses caused by weeds in pulses

Crop	Yield reduction (%)
Kharif Pulses	40-60
Rabi Pulses	20-30

WEED MANAGEMENT

Knowledge of weed biology is essential for the development of weed management systems. Weed problem begins with weed seed in the soil. The weed seeds fall to the ground and remain a source of soil seed bank thereby ensuring perpetual establishment of the weed. Some weeds mature their seeds simultaneously or just before them other crop. Weed seeds remain viable for many years in soil. It is common for weed seeds to retain their viability for ten, twenty or even forty years. Dormancy in weeds includes periodic flushes of weeds. By the time seeds from initial weed populations are eliminated from the soil, many more millions are produced and stored in it as subsequent weed flushcs. In this context an old saying" one year of seeding is seven years of weeding". The vast majority of weed seeds entering the soil come from annual weeds. Only 4% comes from perennial weeds. For effective weed management following measures should be adopted.

Ahlawat *et al.* (1981) reported that weeding increased seed yield of chickpea by 107% and first 4 to 6 weeks were more critical period for weed competition. Any control measure to ensure effective suppression of weeds during this period should result in increased yield. So far, no single method seems to be fully effective and widely adopted to all environments and situations. Financial resources and technical know how capabilities of farmer are important considerations in the choice of the method. Under good management an integrated approach involving chemical, cultural and mechanical methods will probably provide an effective weed control system in chickpea.

A. Cultural methods of weed control

a. Tillage

Tillage methods are still used in many situations as the most effective and economical methods of weed control. Ploughing, cultivating and harrowing make possible weed control before sowing the crop and there are appropriate tools that can do a satisfactory job even after weed emergence. The disc harrow and chisel plough are used for primary tillage operations. New types of minimum tillage implements that cut and incorporate residues and ridge the soil in one operation are advocated for weed control. The younger and smaller the weeds are, the most efficient and economical will be their control.

b. Tillage combined with irrigation

One of the most effective means of weed control is irrigation before sowing with a quantity of water just sufficient to wet the soil of the future root system. Then cultivating the field superficially, before or during sowing will destroy the weeds that have germinated. The crop seeds are sown in moist soil at a depth that is sufficient to ensure germination without additional irrigation. The surface soil dries out, effectively preventing germination of additional weeds.

c. Crop rotation

Crop rotation is defined as a system of growing different kind of crops in recurrent succession of the some land. The help that crop rotation offer in weed control is not always recognized, even though the practice has been recognized.

d. Composting

Composting has direct and indirect benefits for weed control. Most weed seeds in manure and bedding materials are killed by the high temperatures (>50°C) reached in compost piles. Indirectly, composting and other management factors benefit weed control by enhancing soil fertility.

e. Hoeing and hand weeding

It is a method which almost removes all types of weeds to the satisfactory level. Hoeing and hand weeding are probably the oldest agricultural occupations, and the back-breaking work involved still forms a considerable and inevitable part of the human efforts involved in agriculture. Hand pulling, hoeing and tillage are the traditional methods practiced for a long time. These methods are effective when carried out 2 to 3 times at early stages of weed development. Saxena 1977 reported that hand weeding at both 30 and 60 days after sowing was the most effective treatment followed by hand weeding either at 45 or 60 days after sowing.

B. Chemical methods of weed control

Like most of the grain legumes, chickpea is more tolerant to pre-planting chemicals, which prevent the early establishment of seedling from germinating weed seeds. Balyan and Bhan (1984) working at Hisar observed that pre plant incorporation of fluchloralin 1.5 kg. metribuzin 0.75 kg and alachlore 1.5 kg /ha provided grain yield identical to that obtained in hand weeding (twice) treatment.

Panwar *et. al.* (1987) compared different weed management practices and found that all the herbicides resulted in significant decrease in population and dry weight of weeds as compared to weedy check at 60 DAS. However, at 90 days, none of the herbicides was effective. The dry weight of weeds at 60 and 90 DAS with fluchloralin was significantly lower than weedy check during both the years. In second year, only pendimethalin reduced significantly the population and dry weight of *chenopodium album*. Presence of weeds for whole season reduced the potential yield of chickpea by 54 and 69 percent in 1982 and 1983, respectively. In both the years, fluchloralin and pendimethalin produced significantly higher grain yield than weedy check. Similarly, in another study pre plant incorporation of pendimethalin @ 1.0 kg/ha gave effective control of weeds in chick pea (Anonymous, 2000).

Two types of chemicals namely selective and non selective (total killer) are applied either on foliage or on soil and the weeds are killed either through contact or translocated (systemic) action. The method is very effective, quick and convenient but it leaves harmful residues in soil and in plant parts which may affect the succeeding crop and those who consume the produce. The chemicals are applied at pre sowing, pre-emergence (before sprouting of weeds) or post emergence (after germination of weeds) stage according to the nature of herbicides and its killing effect on weeds.

Table 8.2. Weed control chemicals for chickpea.

Name of Crop	Name of herbicide found selective	Rate of application	Time of application of herbicide
Chickpea	Trifluralin	1.5 kg a.i./ha	Pre-emergence
	Prometryne	1.0 kg a.i./ha	-do-

C. Integrated weed Management

Chemical control of weeds alone in chickpea is promising in spite of some technical limitations in its adoption in certain areas. Most of the soil-acting herbicides have limited persistence in the soil and these are only effective at early stage of crop development. The narrow adoption of these herbicides and the inconsistency of their effect from season to season is other limitation. Herbicidal efficacy being highly dependent on soil moisture is bound to vary from one season to another in semi arid areas where rainfall is highly variable. Post emergence herbicides that could effectively control broad leaf weeds satisfactorily are not available till date. Therefore, integrated weed control can be a solution. Study conducted at ARS Durgapura (Jaipur) revealed that pre-emergency application of Pendimethalin @ 0.75 to 1.0 kg/ha could replace one operation of hand-weeding in chickpea to give pulse production equal to the twice hand weeding.

REFERENCES

Ahlawat, I.P.S.; Saraf, C.S. and Singh, A. 1979. *Indian J. Agron.* **24**: 247-249.

Ali, Masood 1989a. Consolidated report of *Kharif* Pulses Agronomy, D.P.R., Kanpur.

Ali, Masood 1989. Consolidated Rep. of *rabi* pulses Agronomy D.P.R. Kanpur (U.P.).

Balyan, R.S. and Bhan, V.M. 1984. Promising herbicides for weed control in chickpea. *Indian J. Weed Sci.* **16(2)**: 69-75.

Balyan, R.S.; Malik, R.K.; Bhan V.M. & Singh, R.P. 1988. Studies on pre- and post emergence weeding systems in mung bean *Indian J. Agron.* **33(3)**: 234-237.

Chopra, Nisha 2004. Advances in weed management. Winter School on "Development of Quality Seeds: Principles and Practices" (8-28 September, 2004). IARI Regional Station, Karnal, Haryana. **PP:** 48-54.

Dayal, Devi. 2004. Integrated weed management in groundnut. Groundnut research in India. Edited by M.S. Basu and N.B.

Singh National Research Centre for Groundnut, PB 5, Junagadh 362001, Gujarat, India. **PP:** 248-259.

Dharkar, P.M.; Pawar, W.S. and Jugele 2000. Economics of weed control in groundnut. *Journal of Soils and Crops*. 10: 307-308.

Giri, A.N.; Bhosle, R.H. and Shelke, D.K. 1998. Weed management in cotton groundnut sequence. *Indian Journal of Agronomy*. 43: 50-56.

Gupta, O.P. 1998. Modern weed management. Agro Botanica. Publishers, Bikaner. India. pp. 1-1J88.

Kundra, H.C. *et al.* 1993. Efficacy of weed management practices in pea *(Pisum sativum* L.) *Indian J. Weed Sci.* **25 (1 & 2)** 1-6.

Mani, V.S.1972. Proc. Eight workshop (Rabi Pulses) ICAR:158-161

Parmar, J.S. *et al.,* 1994. Weed Control Studies in Lentil (*Lens culinaris* M.) *Weed Sci.* **26 (1&2)** 49-50.

Panwar, R.S., Malik, R.K. and Bhan, V.M. 1987. Studies on weed management in wheat–ray and chickpea–ray intercropping. *Indian J. Agron.* **32 (4):** 365-369.

Saxena, M.C. 1977. *Res. Winter Pulses*, GBPUT, Pantngagr Nainital (U.P.).

Saxena, M.C. and Singh, H.P. 1977. Research on *rabi* pulses, GBPUA&T, Pantanagar, Nainital (U.P.).

Saxena, M.C.; Yadav, D.S. and Singh, N.P. 1971. Report of the Agronomic work on *summer pulses* at Pantnagar, Nainital (U.P.).

Saxena, M.C.; Yadav, D.S. and Singh, N.P. 1972. Report of the Agronomic work on *Summer pulses* in 1972 at Pantnagar, Nainital (U.P.).

Sekhawat, M.S. 1988. *Indian J. Agron.* **33 (3)**: 341-42.

Sharma, A.R. and Vats, O.P. 1988. *Indian J. Agron.* **33(2)**:214-215.

Singh, Govindra; Bhan, V.M. & Tripathi, S.S. 1984. *Indian J. Weed Sci.* **XVI (1)** :17-21.

Singh, N.P. and Sharma, B.B. 1988. *Res. Rabi Pulses* GBPUA&T, Pantnagar, Nainital.

Singh, O.P. and Tripathi, S.S. 1984. *Indian J. Weed Sci.* **XVI**:49-51

Singh, V.K. and Bajpai, R.P. 1991. Studies on manual weed control in rain-fed groundnut. *Indian Journal of Agronomy*. 36. 127-129.

Yadav, D.S. 1983. Report of the research work on *rabi* pulses NDUAT, Faizabad (U.P.).

9

Stress Management in Chickpea in Resilience with Climate Change

Surendra Singh, S.S. Shekhawat and O.P. Khedar

Today Climate change is one of the greatest challenges for the world. The semi arid agriculture of Rajasthan is extremely vulnerable to weather and climate (Anonymous, 2007). In recent past there was substantial loss of crop in the region due to extreme and unusual weather conditions. Semi arid eco system has reason to concern about the impact of climate change as its large population depends upon climate sensitive sectors like agriculture and forestry for livelihood. Any adverse impacts of water availability would threaten food security and rural households (Dash and Hunt, 2007).

Thus climate change is a complex alteration of climate, subtle and continuous yet important through its consequences on vegetation of various types that thrived under constant or relatively unchanged Climate. Some of the main effects of climate change with specific reference to agriculture and food production especially during the last decade are: Increased occurrence of drought, frost, flood, heat stress and particularly in Rajasthan state increased frequency of pest and diseases etc. (Venkateshwaralu and Shankar, 2009).

In recent years, the incidence of climatic extreme events like drought, frost, heat stress, flood has become frequent in the country. Climatic changes have already set in, resulting in changes in rainfall and temperature. The analysis of temperature data of the last 100 years indicates a rise of 0.6°C in Rajasthan's temperature (Anonymous, 2007). The analysis further reveals that climate of Rajasthan has exhibited a continuing trend towards desiccation, particularly after the severe drought of 1987. For example, in February and March of 2006, 2008, 2011 and 2012 there was high temperature stress, and there was heavy frost in 2007, 2010, 2011 and 2012 (Anonymous, 2011).

Chickpea cultivation under stress condition

- Chickpea often experience multiple abiotic stresses due to uncertainties in rainfall and extreme weather events
- Drought, heat, salinity and water-logging are the major constraints limiting productivity of chickpea in India.
- Diminishing dew precipitation during winter in northern India

Therefore, chickpea is most vulnerable to climate change. Chickpea is already grown under stressful environments and rising temperature is an additional threat for the sustainable production of gram. Moisture, heat and cold stress are major abiotic stresses experienced by chickpea in recent years which are limiting chickpea production in semi arid ecosystem.

(A) Moisture stress management

Chickpea is largely grown under rainfed condition and suffers badly due to moisture stress in terms of loss in economic productivity. Moisture stress during sowing and at maturity stage is a common phenomenon in northern India. Effect of moisture stress on yield may be assessed in chickpea in terms of yield componants, some of which can be more important than others depending upon intensity of stress and growth stage at which stress develops (Motzo and Deidda, 1993). Reduction in numbers of spikelets per unit area and numbers of kernels per spikelet have been associated with pre-anthesis water stresses (Day and Intalap, 1970). Water deficit at Anthesis may lead to reduced yields by decreasing spikelet numbers and fertility of surviving spikelets (Aspinall, 1984). Water stress during grain filling stage hasten leaf senescene, reduce duration of grain filling and reduce grain weights (Austin, 1989), presumably by reducing assimilate supply to developing kernels. Reduced assimilate supply has been related to reduced photosynthesis (Denmead and Miller, 1976) and early senescence.

Crop management strategies for reducing moisture stress in chickpea are:

a. **Cultural practices:** Hoeing and weeding are most important cultural practice for minimizing water loss from soil by breaking capillary pores. In some areas straw mulch and plastic mulch are also use for moisture conservation.

b. **Planting moisture stress tolerant cultivars**: Some of the cultivars of chickpea are tolerant to moisture stress prominent among them are RSG 888 (Anubhav), RSG-973 (Abha), CSJD-884 (Akash), RSG-895 (Arpita) and RSG-902 (Aruna)

c. **Use of Agrochemicals:** Use of either of the following agro chemical was found to be effective in ameliorating moisture stress in chickpea.

   I. Foliar spray of thiourea @ 500 ppm at pre flowering and pod filling stage gave promising results at Durgapura.

   II. Sprays of 2% urea at flower initiation and pod initiation stage.

   III. Spray of 2% TSP (Triple Super Phosphate) or Potassium Sulphate gave best results at Dholi.

(B) Heat stress Management

A recent simulation study has shown that a rise in temperature by 1°C can lead to a decline in chickpea (gram) yield by 200 kg/ha (Kalra *et al.* 2008). CO_2 enrichment and temperature rise could probably result in depletion of soil water at a rapid during vegetative phase of growth leading to moisture stress during grain filling and consequently poor harvest index and low productivity. In north India warming would reduce some losses in yield by early pod set in winter grain legume like gram (Sharma, 2009). Temperature above 35°C is detrimental for higher productivity of chickpea.

a. Frequent increase in temperature >35°C causes flower abortion and poor pod setting

b. High night temperature adversely affects carbon translocation to grains.

c. Reproductive parts are sensitive and damaged at >39°C

Important adaptive strategies for mitigating the adverse impact of heat stress in chickpea are

1. Planting of heat tolerant genotypes: Vijay (Phule G-1), Avrodhi, ICC 4958
2. Use of agro-chemicals: The experimental study at Durgapura indicated that two foliar spray of Thioglycolic acid (100 ppm) and Salicylic acid (100 ppm at pre flowering and pod formation stags proved significantly superior in increasing the seed yield of gram under heat stress situation in comparison to control and water spray.

(C) Management of cold stress

Black frost occurs when the temperature drops to freezing point (Dew point), but the adjacent air does not contain enough moisture to form white frost on exposed surfaces. This causes an internal freezing of the vegetation, leaving it with a blackened appearance and killing it. The most probable time of occurrence of frost (cold stress) is from mid December to mid Feburary. Cold stress causes abortion of flower and flower drop in chickpea. Followings are important symptoms of cold stress

i. Discolouration and dropping of flowers

ii. Yellowing of young pods and emptiness of pods.

iii. Loss in seed weight

Low minimum temperature (<4°C), calm wind, clear sky and low relative humidity are conducive for cold stress

Technologies for mitigating the adverse impact cold stress

(1) **Use of Chemicals:** Spray of 0.1% H_2SO_4 and Thiourea (500 ppm) is recommended at pre flowering and pod formation stage for mitigating the adverse impact of frost in chickpea.

(2) **Advance planting of chickpea:** Planting of chickpea in the second week of October avoids flowering and pod filling stage to cold stress.

(3) **Cold tolerant genotypes:** RSG-945 (Asha) and RSG-963(Adhar) are tolerant to cold stress.

(4) **Application of light irrigation:** Light irrigation preferably through sprinkler system is effective in mitigating the adverse impact of cold stress. Light irrigation release heat from soil.

(5) **Smoking:** burning of waste farm materials in North western direction forms smoke clouds which prevents heat loss from soil and captures back radiation.

REFERENCES

Anonymous 2007. Climate: Variability and change Background material towards preparation of Adaptation strategy for the state of Rajasthan. Dept. Environment, GOR, Rajasthan Pollution Control Board. **PP** 1.

Anonymous, 2010. Vital agriculture statistics of Rajasthan. Commissionarite of Agriculture. GoR, Pant Krishi Bhavan, Jaipur, **PP** 5-4.

Anonymous, 2011. Consolidated progress report on "Adaptation Strategies for mitigating the adverse impact of climate change in late sown Wheat. **PP** 3-4.

Aspinall, D. 1984. Water deficit and Wheat. **PP.** 91-110. In C.J. Pearson (ed.).Control of Crop Productivity. Sydeny, Australia. Academic Press.

Austin, R.B. 1989. Maximizing crop production in water limited environments. **PP.** 13-25. In: F.W.G. Baker (ed.). Drought Ressistance in Pulses. Lodon. Common. Agric. Bur.

Dash, S.K and Hunt, J.C. 2007. Variability of climate change in India.*Current Science*. 93 (6): 783-88

Day, A.D. and S. Intalap 1970. Some effects of soil moisture on the growth of Chickpea. *Agronomy Journal*. 62: 27-29

Demmead, O.T. and B.D. Millar 1976. Field studies of the conductance of leaveas and transpiration. *Agronomy Journal*. 68: 307-311

Kalra, N, Chakraborty, D, Sharma, A, Rai, H.K., Jolly, M. Chander, S., Ramesh Kumar, P, Bhadraray, S., Barman, D, Miltal, R.B., Mohan Lal and Sehgal, M. 2008. Effect of increasing temperature on yield winter crops in north west India. *Current Science*. 94: 82-88.

Sharma, D.D., 2009. Impact of atmospheric carbon dioxide shift in agriculture. Paper presented in conference on effect of environment change on agriculture held at RCA, Udaipur on 20th Dec., 2009. PP 70-72.

Venkateshwaralu, B. and Shanker, A.K. 2009. Climate change and Agriculture: Adaptation and mitigation strategies. *Indian Journal of Agronomy*. 54 (2): 226-30.

10

Integrated Disease Management in Chickpea

O.P. Sharma, S.S. Shekhawat, O.P. Khedar, Kuldeep Sharma, and Pradeep Singh

Diseases are found to be the most important yield limiting factor in legume cultivation. Sometimes diseases occur in epidemic form which results in total failure of the crop. High yielding cultivars are often susceptible to one or more diseases. The basic need therefore, is availability of good quality healthy seed. Seeds are known to be the carriers of a large number of microorganisms. Healthy seed can be obtained through appropriate certification or by effective seed treatments, but it is difficult to restrict the seed transmitted diseases, which are also soil borne or perpetuate on plant residues. In general, chemicals are used to reduce yield losses due to diseases and now a day the indiscriminate use of chemicals has led to the problems of residues. Heavy use of chemicals is also harmful to the eco-system or in other words it is harmful to environment, soil, water, soil microbes and living beings. It has decreased the soil productivity up to great extent. Regular and continuous use of chemical kills the beneficial microbes present in the soil, which are essential for good soil health. By the regular use of chemicals the budget on plant protection has increase in a big way which is dangerous sign for a country like India. An integrated approach is needed for the effective control of the disease and production of pathogen free seed.

To avoid the yield losses, there is an urgent need to develop efficient integrated disease management module involving cultural practices, appropriate cropping system, resistant/tolerant varieties, judicious use of chemicals and biocontrol agents etc. Though, there are a number of diseases in every crop but important diseases are limited in a particular agro-climatic condition so care should be taken to control the prevailing diseases of the area first. It will provide sustainable plant protection for a particular agro-climatic condition. Though, till date it is quite impossible to stop the use of chemicals but care should be taken that it should be used in prescribed dose only and over use should be discarded. As far as legumes are concerned IDM is the only solution because legumes in general are poor yielder due to cultivation on marginal lands in rainfed situation and farmers are not able to invest more on plant protection.

ADVANTAGES OF INTEGRATED DISEASE MANAGEMENT

It is eco-friendly and helpful to overcome soil, water and environmental pollution. It is a simple and easily manageable and helpful in balancing living organisms. It increases soil fertility and productivity. It helps in preserving beneficial microorganisms. It is economic and chances of residual effect are less. It improves quality of the produce, increases storability of the produce and germination % of the seed.

INTEGRATED DISEASE MANAGEMENT STRATEGIES

Chickpea is more sensitive to diseases in comparison to cereals, so for better production, plant protection measures should be adopted well in time because delayed adoption of plant protection measures may result heavy losses to the crop. It is also fact that one method alone cannot control the diseases to the desired level in the field. Therefore, integrated plant protection measures involving cultural practices, resistant/tolerant cultivars, judicious use of chemicals and biocontrol agents may provide sustainable and durable plant protection from diseases.

Integrated disease management strategy combines genetical, cultural, chemical and biological methods in a compatible manner resulting in the growth of healthy plants. Diseases cause losses from seedling to fruiting stage. For effective management of diseases surveillance, monitoring, disease severity data in different areas and meterological data are very important. It is also essential to know the nature of the causal organism, its mode of penetration, method of spread and survival. It is comparatively easy to assess seeds for the associated pathogen and also to control them effectively. Seed health test has been considered as an important component of disease management strategy. A variety of approaches have been found effective in controlling diseases and improving crop production. Judicious combination of different practices suitable for location specific may be adopted for disease management. Indiscriminate use of inputs as fertilizers, pesticides, water or the cropping systems for increased productivity has already created problems of sick soils; lowering of water table, build up of inoculums threshold levels and resistance in pathogens and pesticide residues are all harmful for sustaining agriculture. Hence, there is an obvious need to evolve eco-friendly and environmentally safe IDM technology utilizing less harmful biological and cultural methods of disease control to ensure proper and judicious use of chemical pesticides for sustaining healthy agriculture for the future. The methods of controlling plant diseases are:

A. Monitoring

It provides knowledge about the current status of disease and crop situation and is helpful in selecting the best possible combinations of the disease management methods. It is also an important aspect before adopting any plant protection measure. Prevailing diseases and their intensity, season, climate, soil type, crop etc. are some of the aspects which should be kept in mind before adopting disease management measure.

B. Exclusion of inoculum

It aims to prevent new pathogens and diseases from invading uninfected areas and to avoid contact between the pathogen and the crop. Important exclusion methods to prevent spread of the pathogens are:

a. Resistant varieties

Resistant varieties are eco-friendly and most effective tool to control diseases. It eliminates environmental pollution and chemical residue effects thus protecting mankind and helping in maintaining biological balance in the environment. It is economical and no additional investment is required for crop protection. It does not require any special technical guidance and equipment for its use. However, initially perfect screening and testing is must for the development of resistant material. Though, there are so many genotypes available, even then, it is not up to mark till date and needs more perfection. Disease resistant varieties are short lived as the varietal resistance breaks down due to development of more virulent strains of the pathogen. Though, it is not possible to control all the diseases, nematodes and insect-pests by a single variety but if a particular disease or insect-pest is very much common in a particular area then we may use a variety resistant against them in that area. Resistant varieties are the cheapest tool for plant protection with no residual effect and human and cattle casualties.

b. Quarantine Regulations

The quarantine laws first came in existence in USA in 1912 and are known as Federal Quarantine Acts. In India, there are 26 plant quarantine and fumigation stations (PQ & FS) located at 10 international airports, 9 sea ports and 7 land frontiers. The quarantine responsibility for plant genetic resources has been entrusted upon National Bureau of Plant Genetic Resources, New Delhi for agri-horticultural crops. Viruses such as bean yellow mosaic virus, cowpea aphid borne mosaic virus, pea seed borne mosaic virus, soybean mosaic virus, etc. have also been intercepted on imported legume seeds. Hence, it has become mandatory that the acquired germplasm accessions be grown in post entry quarantine net houses under constant vigil.

c. *Enforcement of seed certification standards*

Seed health certification is an integral component of seed certification programme and is being used in integrated disease management. Seed health certification is a legalized scientific process to secure, maintain, multiply and make available quality seed to the consumers. It is the prerogative of the government to lay down the procedures of certification and fix a minimum permissible limit for a particular disease. This limit may vary from zero to any number depending upon the virulence of the pathogen and the host. It involves testing of seeds before sowing and after harvest, and crop inspection for compliance of standards including isolation and freedom from weeds and diseases. The International Seed Testing Association (ISTA) is engaged in the development of standardized seed testing procedures that are available in the form of International rules for seed testing. In India, Central Seed Certification Board (CSCB) is responsible for framing the rules and procedures for seed certification including seed health. CSCB has fixed the permissible limit for few seed borne diseases on the basis of field inspections or seed evaluations. However, under the provisions of Seed Act 1966, labeling is compulsory but certification is voluntary.

C. Agronomic management/Cultural control

Agronomic management or cultural practices are an effective tool in controlling diseases below economic thresh-hold. The efficacy of cultural practices depends on soil type, disease and its intensity and agro-climatic condition of the soil. It includes crop production practices that make crop environment less susceptible to the pathogens.

Summer fellowing and ploughing reduces the incidence of diseases. Deep ploughing exposes the spores of diseases and the spores are destroyed by sunlight up to some extent. Crop rotation is also an effective tool for reducing the population of disease causing organisms. Proper field preparation and tillage operation reduces weeds, which are alternate host for disease causing organisms. It also improves the soil texture, structure and pH which does not favour diseases. Proper fertility management helps in reducing the attack of diseases because imbalance of nutrient changes the chemical composition of the plant favors the attack of diseases. Cultural controls are based on pest biology and development. Before adopting any cultural practice it is essential to know the biology of disease, host range, soil type and agro-climatic condition.

a. Summer fellowing and ploughing

Summer fallowing and 2-3 deep ploughing helps in reducing diseases and improves plant growth and yield too. It helps in destruction of spores of diseases when exposed to the sunlight.

b. Crop rotation

Most of the diseases have wide but specific range of hosts so try to accommodate commercially acceptable and agronomically suitable non-host or poor host crop in crop rotation. Legumes may be cultivated in rotation with cereals, millets or grasses. Long term crop rotation is more effective in controlling diseases.

c. Proper field preparation

Proper field preparation and tillage operation reduces weeds, which are alternate host for diseases. It also improves the soil texture, structure and pH, which does not favour diseases.

d. Fertility management

Proper fertility management helps in reducing the attack of diseases because imbalance of nutrient changes the chemical composition of the plant, which favors the attack of disease.

e. Inter-cropping

Some of crop species showed antagonistic effect to the particular disease. Inter-cropping of legumes with cereals helps in reducing disease infestation, however, appropriate crop geometry is essential to avoid shading effect.

f. Cropping system

Cropping system also play an important role in disease manage-ment. Continuous cultivation of the same crop in the same field (mono-cropping) leads to build up of inoculum levels of the pathogen in the soils thus making them sick. Proper crop rotation helps to control many soil borne diseases as wilt of arhar, pea, gram etc. Crop rotation also helps to replenish the soils with essential nutrients. If host crops (Vegetables) are grown prior to legumes results in heavy infestation of diseases. If non-host crops are grown prior to legumes reduce the disease infestation. Fellowing also helps in reducing diseases below economic threshold level.

g. Avoid mono-cropping

Mono-cropping helps in build-up of disease inoculam so avoid mono-cropping.

h. Manipulation in sowing dates

The time of sowing also influence the incidence of diseases. Pathogens are able to infect susceptible plants only under certain environmental conditions e.g. root rot caused by *Rhizoctonia* is severe if gram is sown immediately after rains. The early maturing pea varieties also escape damage due to *Erysiphe polygoni*. Most of the diseases occur at a particular time so early or late planting is also an effective measure for disease management.

i. Weed free cultivation

Weed free cultivation helps in drastic reduction of diseases because some of the weeds act as host for disease. Clean cultivation helps in disease management up to a remarkable level.

j. Organic amendments

Incorporation of vermicompost or FYM into soil reduces disease intensity. Application of non-edible de-oiled seed cake of neem improves soil fertility and reduces disease intensity.

k. Selection of disease free plot for healthy seed production

Selection of suitable area or field for cultivation of crops for better yields and also for protection against diseases is essential. Disease free seed production is planned in safe areas and seasons where disease development is restricted or absent. In case of soil borne pathogens, it is advisable not to use the same field for cultivation of crop year after year e.g. wilt of arhar. The drainage in field is also important. Low lying, waterlogged conditions favour diseases.

D. Eradication of pathogen

These methods help to reduce the source of infection in an established disease primarily by breaking the disease cycle. To do so, it is essential to know the life cycle of pathogen, its mode of penetration, host range and growth habit for devising control strategy for a disease.

a. Roguing of infected plants at proper time

The infected plants or plant parts are carefully removed from the fields and destructed at an early stage to remove foci of infection and restrict spread of the pathogen. Roguing of diseased plants as early as possible is essential to prevent the spread of infection. To control bacterial and viral diseases, up-rooting and destroying the diseased plants are the best strategy.

b. Eradication of Alternate or collateral hosts

Many pathogens require two hosts for completing their life cycles. The destruction of one wild host breaks the life cycle chain.

c. Field sanitation

Field sanitation is also important for control of many diseases. Pathogens survive through dormant structures in the soils. Hence, it is essential to collect and destroy all these structures and diseased stubbles. The soils infested with pathogen should not be moved from one area to another and fumigation should be practiced. If possible, allow fallow period between two successive crops especially during hot summers when the land is exposed to hot sun after ploughing. Solarization helps disease management. Solarization by covering the soil with transparent polyethylene sheets during the summers (April, May and June) significantly reduces the spores of diseases. Irrigation before solarization enhances the effectiveness of solarization. Solarization also helps in reducing soil borne diseases, nematodes and other pests.

d. Seed sanitation

Use of pathogen free seed material is the most effective method for disease free seed production. The inoculums present as external contaminant e.g. spores, mycelium, or sclerotia can be removed from the seed lot by physical methods as sieving, steeping in water, or brine solution. Heat therapy can be used for eradication of internal infections in the seed lots. Seed should be dried well before store.

E. Seed treatment

Seed treatment is also cheaper and effective way of disease management. Seed treatment helps in the elimination of seed-borne pathogens that cause seed rots, seedling blights, and also protects the germinating seed from the attack of soil borne pathogens. Seed treatment is a simple, cheapest and safest method of controlling different pests. Seed is treated to promote good seedling establishment, minimize yield losses, maintain and improve the quality of seed and to avoid the spread of harmful organism. Different seed treatment methods have been proposed time to time for control of the disease. It provides initial protection to the young plants in the early stage. To overcome the seed-borne diseases, seeds should be treated with Bavistin or Thiram or Captan @ 2, 3, 2.5 g/kg of seed respectively. Captan and thiram are extensively used as seed dressing fungicides against damping off of beans and seedling rot of groundnut.

F. Chemical control

Chemical control found to give quick and effective control of diseases. Though, high cost, environmental hazards, residual effects, unsafe for human and cattles etc. are some of the limitations in its use. It also requires technical knowledge to use and minor mistake results in heavy loss of crop and human health. In general the use of chemicals is economical only in high value crops but if severity is more, there is no option other than chemical control. Foliar application of chemicals protects the plants against the pathogens which attack the growing plants. These sprays can be preventive or curative. Depending on the type of action these chemicals can be surface protectants or systemic in nature. The protectant pesticides are prophylactic in their behaviour and form a protective barrier on the surface of the seed/plant system that discourages the growth of the pathogen. The systemic pesticides penetrate the plant tissues and kill the pathogen already established in the tissues. The chemical pesticides, especially the fungicides, are of high value in plant protection but the problem of residue build up in the produce. The use of chemicals for the protection of plants against diseases can be integrated with other strategies for disease management that do not have associated health risks.

H. Different diseases their symptoms and control measures

Wilt, dry root rot, *Aschochyta* blight, botrytis grey-moulds, collar rot, stunt and stem rot are the diseases of economic importance in chickpea. The causal organism, symptoms and integrated approach for control of important diseases in chickpea are as under:

a. Fungal diseases affecting arial plant parts

Stem rot

Causal organism: Sclerotinia sclerotiorum (Lib.) de Bary

Symptoms: It is a seed–borne fungal disease appears on adult plants, especially crop with thick canopy and excess moisture. The disease is characterized by the appearance of chloratic or drying branches or whole plants scattered in the fields. Such drying plants or branches rot at the collar region or at any point on the branch. The petioles and leaflets of affected plants turn yellow or droop, while remaining green then dry up and turn straw coloured prematurely. A web of white mycelial strands appears in the color region and above.

Control measures:

1. Avoid excess moisture.
2. Clean cultivation.

Ascochyta blight

It is major disease in West Asia, Northern Africa and Southern Europe. The disease usually built in February March in Pakistan and Northern India when the crop canopy is very dense and temperatures are favorable for disease development. The disease builts up and spreads fast when night temperatures are around 10°C, day temperatures are around 20°C and rains are accompanied by cloudy days. The disease is both externally and internally seed borne.

Causal organism: Ascochyta rabiei (Pass.) Labr.

Symptoms: Small circular spots with brown margins and grey center that contains pycnidia that are often arranged in concentric rings. On the stem and petioles, the lesions are obvate or elongate and bear pycnidia. The size of lesions varies greatly. The stems and petioles usually break at the point of gridling. If blight occurs at pre-flowering stage and then conditions for its development become unfavorable, the crop re-grows fast but symptoms can still be seen on the older branches. Fully developed lesions on pods are usually round up to 0.5 cm in diameter, usually with concentric rings of pycnidia. If infection occurs at early stage several lesions may appear on a single pod.

Control measures

1. Seed treatment with Bavistin + Thiram (1:2) or Captan @ 3 g/kg of seed.
2. Sowing of disease free seed.
3. Removal and destruction of diseased plant debris.
4. Deep ploughing.
5. Intercropping with wheat or barley.
6. Use of resistant (BG 261, C 235, GNG 146, GNG 469, GNG 663, PBG 1 etc.) or tolerant (BG 256) varieties for sowing.

Botrytis grey mould

It is serious disease of parts of Bangladesh, India, Nepal and Pakistan. It is a fungal disease most common in tarai region. It is seed-borne disease and appears regularly in moderate to severe form. Cool weather and high humidity favors the disease.

Causal organism: Botrytis cinerea Pers. Fr.

Symptoms: The disease is usually seen at flowering time when crop canopy is fully developed. Lack of pod setting is the first indication of the disease. Leaves and stems may not show any symptoms. Shed flowers and leaves covered with the spore mass, can be seen on the ground under the plants. When the humidity is very high symptoms appear on stems, leaves, flowers and pods as grey and dark brown lesions covered with moldy sporophores. Tender branches break off at the point where the grey mould has caused rotting.

Control measures

1. Lower seed rate with wider spacing.
2. Late planting in the first fortnight of November particularly in irrigated areas.
3. Adopt crop rotation.
4. Seed treatment with Bavistin + Thiram (1:2), or Bavistin @ 3 g/kg of seed.
5. Spray of Captan (0.2%) 3-4 times at an interval of 15 days.
6. Use of resistant varieties [{Sweta (ICCV 2) Kabuli}, SAKI 9516 etc.] for sowing.

b. Fungal diseases affecting root or stem base

Wilt

Wilt, an important disease of chickpea widely prevalent in almost all regions of the country. The estimated loss by this disease is near about 10%.

Causal organism: Fusarium oxysporium Schlecht. emen Snyd. & Hans. f. sp. *ciceri* (Padwick) synd. & Hans.

Symptoms: It is a seed-borne as well as soil-borne fungal disease. The pathogen can survive in the soil up to six years even in the absence of the host. It is a typical vascular disease causing xylem necrosis. The disease is systemic and plants may be infected at any stage. It generally appears after 3 weeks of sowing. The field symptoms of wilt are dead seedlings or adult plants usually in patches. Dropping starts from the upper part of the plant but soon the entire plant drop down. Whole seedlings collapse and lie flat on the ground. Sometimes only partial wilting occurs. These seed lings retain their dull green colour. When up-rooted they usually show uneven shrinking of the stem above and below the collar region. The shrunken portion may be about 2.5 cm or longer. Affected seedlings do not rot on the stem or root surface. However, when split open vertically from the collar downwards or cut transversely, dark brown to black discoloration of the internal stem tissues is clearly visible.

Control measures

1. Avoid cultivation in heavily infested fields.
2. Inter cropping or crop rotation with wheat, barley, linseed and mustard.
3. Deep ploughing in summers.
4. Soil solarization by covering with transparent polythene sheet for 6-8 weeks during April-May.
5. Removal of host plant debris from the field.
6. Rouging of diseased plants.
7. Seed treatment by Bavistin + Thiram (1:2) @ 3g/kg of seed or Bavistin (Carbendazim) @ 2 g/kg of seed or *Trichoderma viride*+Vitavex @ (4:1) 5 g/kg seed.
8. Use of resistant/TOLERANT varieties for sowing like: Annegiri 1 (Kabuli), Avarodhi, BG 372, BGD 72, DCP 92-3, GNG 469 (Samrat), Gujarat Gram 1,

Haryana Chana 1, JG 11 (*Kabuli*), JG 74, JG 315, KPG 59, Phule G 5, GNG 1581, Pusa 212, ICCC 32, Vishwas, Vijay, Vishal, GF 89-36, KWR 108, KGD 1168, PBG 1, GNG 663, GPF 2, ICCV 10, RSG 888, Phule G 95311, RSG 963, Kranti, Virat, JG 130 etc.

Dry root rot

It is most important disease of chickpea prevalent in most of chickpea growing countries. Generally disease appears around flowering and podding stage. The susceptibility of plant to the disease increases with age.

Causal organism: Rhizoctonia bataticola (Taub.) Butler

Symptoms: It is a soil–borne fungal disease. Maximum temperature above 30°C, minimum above 20°C and moisture stress is favorable for the disease. Disease appears in scattered patches. The petioles and leaflets droop only at very top. Sometimes when rest of the plant is dry, the topmost leaves are chloratic. The leaves and stem are usually straw colored, but in some cases the leaves and stems are brown. Infected plants dry up. When plants pulled out root remains in soil. It turns dark showing sign of rotting. Dark minute sclerotial bodies can be seen on the root exposed or inside the wood. The tap root is dark, shows signs of rotting and is devoid of most of its lateral and finer roots.

Control measures

1. Use of resistant varieties like: (Moderately resistant RSG 888, RSG 931, RSG 807, RSG 963, RSG 974, RSG 898, GNG 1581 etc.)
2. Adopt crop rotation.
3. Seed treatment by Carbendazim + Thiram (1:2) 3 g/kg of seed or *Trichoderma viride*+Vitavex @ (4:1) 5 g/kg seed.

Collar rot

Causal organism: Sclerotium rolfsii (Sacc.) Curzi.

Symptoms: High

It is a soil–borne fungal disease mostly appears in seedling stage when soil moisture is high and temperatures are warmer (30°C) at sowing time. The incidence decreases with the age of the crop. Drying plants whose foliage turns slightly yellow before death, scattered throughout the field is an indication of collar rot infection. Most often, collar rot is seen at seedling stage particularly if the soil is wet. Affected seedlings turn yellow. Young seedlings may collapse, but older seedlings may dry without collapsing. Rotting of stem can be seen at the color region, the rotten portion of the stem covered with whitish mycelial growth of the fungus and disease may appear in patches.

Control measures

1. Seed treatment by Captan @ 2.5 or Thiram @ 3g/kg of seed.
2. Crop rotation.
3. Avoid excess soil moisture.
4. Drenching of Demosan @ 15 kg/ha.

Black root rot

Causal organism: Fusarium solani (Mart.) Sacc.

Symptoms: It is a seed–borne fungal disease generally occurs under high moisture condition and moderately high temperatures (25-30°C) at any growth stage. It is minor disease reported from Argentina, Chile, India, Mexico, Spain, Syria and USA. Affected plants turn yellow and wilt. Dead plants are seen scattered in the field. The root system is rotten. Most of the finer roots are shed, and the remaining roots turn black.

Central portion of stem affected by fusarium wilt

Flowers killed by Botrytis gray mold

Root-knot nematode

Black root rot : lesions above the collar region (left); symptoms on roots (right)

Ascochyta blight

Ascochyta blight

Stunt : Discolored leaflets (left); browning of phloem (right)

Control measures

1. Seed treatment by Carbendazim+Thiram (2+1) 3 or Captan @ 2.5 or Thiram @ 3 g/kg of seed.
2. Proper drainage.
3. Roguing of infested plants.
4. Use of resistant varieties.

c. Viral disease

Stunt

This is the most important viral disease of chickpea prevalent in most of the chickpea growing countries. In India it is prevalent in the state of Gujarat. All other viral diseases are minor.

Causal organism: Chickpea chloratic dwarf virus, bean leaf roll virus.

Symptoms: Stunting is the most conspicuous in early infections. It occurs because of shortened internods. In later infections stunting may not be obvious but plant discoloration and phloem browning are seen. Affected plants can be easily spotted in the field by their yellow, orange, or brown discoloration and stunted growth. The disease is not seed-borne. The tips and margins of leaflets often become chlorātic before turning reddish-brown. In general leaf discoloration in more pronounced in desi types than in kabuli types. The leaves of diseased plants are stiffer and thicker than normal. The most characteristic symptom is pholeum browning. It becomes obvious if the bark is removed at color region (by cutting a thin slice lengthwise) a transverse cut reveals a brown ring or a split through the color region reveals brown streaks of discolored phloem vessels.

Control measures

1. Use resistant varieties for sowing like (RSG 888).
2. Control of leafhoppers and aphids.
3. Late sowing.

G. Biological control

Bio-control agent has become an emerging trend in the area of integrated diseases management. These include augmentation and conservation of natural enemies of pathogens such as use of *Trichoderma.* Bio-products based on plant origin and bio-pesticides have shown promise against disease management. Their easy availability, rapid biodegradation, low toxicity, self perpetuation, compatibility with other products, ecologically and environmentally safe properties make them more reliable and reduce the usage and dependence on chemical pesticides. However, prior testing of such organisms to human and cattle health is must. They control the pathogens through antagonistic activity of microorganisms, which is achieved by parasitism, predation, competition or commensalisms.

COMPATIBILITY OF FUNGICIDES AND OTHER AGROCHEMICALS

Some-times more than one chemical are used for crop production so before using any chemical care should be taken that those chemicals must be compatible to each other. It should be taken care that the chemical used should not affect the effect of previous chemical. For example: Captan may be used with any insecticide, fungicide or any organic manure but it decomposes with alkaline material. Some of the agrochemicals help in increasing the efficacy of the fungicides like:

i. Copper oxy chloride with Zinabe, captan, qunolinol and sulphur.

ii. Surphur with Zinabe.

iii. Zinabe with Thiram, Captan sulphur and minabe,

Care should be taken that chemicals are effective in a particular ratio only.

REFERENCES

Gupta, Anuja 2004. Integrated management of diseases in seed crops. Winter School on "Development of Quality Seeds: Principles and Practices" (8-28 September, 2004). IARI Regional Station, Karnal, Haryana. **PP:** 116-122.

Gupta, V.K. & Sharma, R.C. 1995. Integrated disease management and plant health, Scientific Publishers (India), Jodhpur.

Kaur, Livinder. 2007. Pulses disease scenario in Punjab. Pulses at a glance. Pulses section, Department of Plant Breeding, Genetics and Biotechnology, PAU Ludhiana.

Khare, M.N.; Kumar, Kumud and Gurha, S.N. 2000. Management of Seed Borne Pathogens of Pulse Crops. Advances in Management of Biotic and Abiotic Stresses in Pulse Crops. Proceedings of National Symposium on "Management of Biotic and Abiotic Stresses in Pulse Crops" held at Kanpur, June 26-28, 1998. **PP:** 27-33.

Khedar, O.P.; Singh, R.V.; Shrimali, Mahesh and Singh, N.P. 2008. Pulses: Status and Cultivation Technology. Aaviskar Publishers, Distributors, Jaipur.

Nene, Y.L. Reddy, M.V. Haware, M.P. Ghanekar, A.M. and Amin K.S. 1991. Field diagnosis of chickpea diseases and their control. ICRISAT Information Bulletin No. 28. Patancheru, AP.

Singh, R.A.; Gurha, S.N. and Chand, Hari 2000. Integrated management of fungal foliar diseases in mungbean. Advances in Management of Biotic and Abiotic Stresses in Pulse Crops. Proceedings of National Symposium on "Management of Biotic and Abiotic Stresses in Pulse Crops" held at Kanpur, June 26-28, 1998. **PP:** 99-103.

Tanwar, N.S. and Singh, S.V. 1988. Indian Minimum Seed Certification Standards. The Central Seed Certification Board, Department of Agriculture and Co-operation, Ministry of Agriculture, Government of India, New Delhi, **PP:** 388.

Upadhyay, J.P.; Lal, H.C. and Ojha, K.L. 2000. Biological Control of Soil Borne Pathogens in Pulse Crops. Advances in Management of Biotic and Abiotic Stresses in Pulse Crops. Proceedings of National Symposium on "Management of Biotic and Abiotic Stresses in Pulse Crops" held at Kanpur, June 26-28, 1998. **PP:** 21-25.

Vishwa Dhar, Singh, R.A. and Gurha, S.N. 2003. Integrated Disease Management in Pulse Crops. Pulses in New Perspective. Proceedings of National Symposium on Crop Diversification and National Resource Management. 20-22 December 2003 IIRP, Kanpur.

11

Integrated Insect Pest Management in Chickpea

R.S. Mahla, S.S. Shekhawat, S.R. Dhaka and A.S. Baloda

In general the productivity of chickpea is very low and unstable, primarily due to biotic and abiotic stress. Among the biotic stress insect-pests are found to be the most important biotic stress and yield limiting factor. Insects affect plant stand, health, seed setting and its quality and also damage the stored seeds. To avoid the yield losses, proper plant protection measures should be adopted well in time, because delayed or improper plant protection measures may adversely affect the production, quality and storability of the seed. Indiscriminate use of insecticides has lead to the problems of residues, resurgence and resistance in insects. Again it is also fact that one method alone cannot control insect-pests to the desired level. Therefore, there is an urgent need to develop efficient integrated pest management module involving cultural practices, appropriate cropping system, resistant/tolerant varieties, judicious use of chemicals and bio-control agents.

WHY CROP PROTECTION

Now a days, crop protection attains prime importance and is an essential component in crop production because without crop protection there are chances of total failure of the crop which ultimately hamper the seed chain and production in the coming years. Some of the reasons why crop protection is essential are listed here:

A. India with diversified agro-ecosystems responded spontaneously to the technologies of green revolution with introduction of several components in crop production like developing and adopting high yielding varieties, hybrids, usage of new agro-chemicals and adoption of intensive crop cultivation techniques.

B. The gains of green revolution reflected in the shape of production of food grains and oil seeds. But these steady gains in agricultural production have not fully overcome the problem of rising demand caused by soaring population growth.

C. Adding to the population explosion, there were frequent setbacks to crop production experienced in the shape of abiotic and biotic stresses during the last two decades in several food crops where intensive farm practices were adopted.

D. Among these stresses on major crops, increased pest populations leading to the stage of collapse of economy, at times keep the planners and executors to be helpless. In the past one and half decade, the periodical unabated explosions of aphids, whiteflies, pod borers, defoliators, cutworms, plant hoppers etc., as direct crop damagers and disease transmitters in different regions of the country have made agriculture less remunerative and highly risk prone.

E. The ability of some of these pests to develop resistance, curbs the effectiveness of many commercial chemicals. Resistance has accelerated in many insect species and it was reported that more than 500 insect and mite species are immune to one or more insecticides at present. Similarly about 150 plant pathogens such as fungus and bacteria are now shielded against fungicides. Some of the weedicides also found effective earlier failed to control weeds now-a-days.

F. Hence, there is need to reduce if not eliminate these losses by protecting the crops from different pests through appropriate techniques. At present day the role of crop protection in agriculture is of great importance and a challenging process than before, as the so called resistant species should be brought under check.

G. All other management practices of crop husbandry will be futile if the crop is not protected against the ravages of pests. In absence of crop protection the yields may be drastically declined. The entire effort of growing a crop will be defeated in absence of crop protection resulting in financial loss to the grower. So the crop protection against various pests is a must in agriculture.

INSECT-PESTS AND THEIR MANAGEMENT

'PEST' is an organism that causes damage resulting in economic loss to a plant or animal. It can also be said that pest is a living organism that thrives at the expense of other living organism. The main insect-pests of chickpea are: Pod borer (*Helicoverpa armigera* Hubner), black aphid (*Aphis craccivora* Koch), and termite (*Odontotermes obesus* Ramb and *Microtermes obesi* Heomgr). Pod borer is a key pest causing 90-95% damage alone. It is very difficult to manage insect-pests through one control measure or in other words one method alone cannot control insect-pests to the desired level. IPM is a system that in the context of the associated environment and the population dynamics of the pest species utilizes all suitable techniques and methods in as compatible manner as possible and maintains the pest populations at levels below those causing economic injury (FAO, 1972). In integrated pest management both crop and pest are seen as part of a dynamic agro-ecosystem.

Integrated Pest Management Strategy

1. While developing IPM strategy one has to select different components that are readily available, economical and applicable at field level.
2. To cater the needs for location specific cropping systems the suitable technologies should be developed by research workers from time to time.
3. The research findings that are practically implementable should be popularized by the extension workers through education to farming community. Farmers have to be trained in scouting, diagnosis of pest infestation and arriving ETLs for need based chemical application in time.
4. Farmers should also be trained in selection of suitable pesticide, use of proper lethal dose and proper coverage of foliage to avoid risks of resistance, resurgence and residues. Farmer's training is a continuous process and is an important integrated part for successful implementation of IPM.
5. A healthy, meaningful co-operation is very much needed from corporate pesticide industry to make IPM successful at farmer's level. The pesticide industry should not wield enormous financial power and maintain market dominance against ecological and environmental safety.

Advantages of Integrated Pest Management

1. It is eco-friendly.
2. Helpful to overcome soil, water and environmental pollution.
3. Helpful in balancing living organisms.

4. It is a simple and easily manageable.
5. Helps in preserving beneficial micro-organisms.
6. Increases soil fertility and productivity.
7. More economic.
8. Chances of residual effect are less.
9. Improves quality of the produce.
10. Increases storability of the produce.
11. Germination % of the seed is more.

Management of Insect-Pests

A variety of approaches based on cultural, physical, chemical and biological methods have been found effective in controlling insect-pests and improving crop production. Judicious combination of different practices suitable for location specific may be adopted for insect-pest management. The present pest control approach of over kill through pesticide should be avoided. Though, chemicals are helpful in controlling insect-pests and increasing the yield but at the same time they lead to several environmental problems such as development of resistance in the pests for that particular pesticide, destruction of beneficial microorganisms and harmful residual effect on soil, environment and living beings. Therefore, strategy should be worked out for IPM which is economically viable and eco-friendly. Special emphasis should be given to those pests which occur consistently and cause heavy loss to the crops. A number of IPM approaches are there which are very much effective and may be used for pest control. For effective insect-pest management we have to adopt following control measures approaches:

A. Survey surveillance of insect-pests

Survey and surveillance is an important aspect before adopting any plant protection measure. Prevailing insect-pests and their intensity, season, climate, soil type, crop etc. are some of the aspects which should be kept in mind before adopting disease management measure. A continues and vigorous pest surveillance is essential for timely prevention of sudden out breaks of epidemics. If control measure is decided on the basis of the survey then only it will be effective.

B. Physical management

Solarization by covering the soil with transparent polyethylene sheets during the summers (April, May and June) significantly reduces the insect-pests living in the soil. Irrigation before solarization enhances the effectiveness of solarization.

C. Agronomic management

Agronomic management or cultural practices are an effective tool for plant protection. The efficacy of cultural practices depends on soil type, disease and its intensity and agro-climatic condition of the soil. Summer ploughing, crop rotation and inter-cropping are some of the important practices by which insect-pests can be managed easily. Before adopting any cultural practice it is essential to know the biology of insect-pests, host range, soil type and agro-climatic condition.

a. **Summer fallowing and ploughing**: Summer fallowing and 2-3 deep ploughing reduces the incidence of insect-pests and improves plant growth and yield too. Deep ploughing exposes the insects and their eggs which are eaten up by birds.

b. **Crop rotation:** Crop rotation is also an effective tool for reducing the population of insect-pests by breaking regular host cycle. Most of the diseases have wide but specific range of hosts so try to accommodate commercially acceptable and agronomically suitable non-host or poor host crop in crop rotation. Legumes may be cultivated in rotation with cereals, millets or grasses. Long term crop rotation is more effective in controlling insect-pests.

c. **Inter-cropping**: Some of crop species showed antagonistic effect to the particular disease. Inter-cropping of legumes with cereals helps in reducing insect-pests infestation, however, appropriate crop geometry is essential to avoid shading effect.

d. **Proper field preparation**: Proper field preparation and tillage operation reduces weeds, which are alternate host for insect-pests. It also improves the soil texture, structure and pH, which does not favour insect-pests.

e. **Fertility management:** Proper fertility management helps in reducing the attack of insect-pests because imbalance of nutrient changes the chemical composition of the plant, which favors the attack of insect-pests.

f. **Cropping system**: Cropping system also play an important role in insect-pests management. If host crops (Vegetables) are grown prior to legumes results in heavy infestation of insect-pests. If non-host crops are grown prior to legumes reduce the disease infestation. Fellowing also helps in reducing insect-pests below economic thresh hold level.

g. **Avoid monocropping:** Monocropping helps to build-up of insect-pest population greatly so avoid monocropping.

h. **Manipulation in sowing dates:** Most of the insect-pests occur at a particular time so early or late planting is also an effective measure for insect-pests management.

i. **Weed free cultivation**: Weed free cultivation helps in drastic reduction of insect-pests because some of the weeds act as host for insect-pests. Clean cultivation helps in insect-pests management up to a remarkable level.

j. **Organic amendments:** Incorporation of vermi-compost or FYM into soil reduces insect-pest intensity. Application of non-edible deoiled seed cake of neem improves soil fertility and reduces insect-pest intensity.

D. Biological management

Bio-control agent has become important now-a-days. These are economic and safe to human health, however, prior testing of such organisms to human and cattle health is must.

E. Management through plant-host resistance

Resistant varieties are an effective tool for insect-pest management. It is also a cheapest input in insect-pest management and also helps in lowering down the population of insect-pests. Resistance varieties are eco-friendly and may be grown successfully without any damage by insect-pests. It does not require any special technical guidance for its use. It also economical and no additional investment is required for crop protection. However, initially perfect screening and testing is essential for the development of resistant material. Though, there are so many genotypes available in legumes but even then it is not up to mark till date and needs more perfection.

F. Seed treatment

Seed treatment is also cheaper and effective way of disease management. It is quite effective to control early insect-pests. It provides initial protection to the young plants in the early stage. It is cheapest control measure of insect-pests.

G. Chemical management

Chemical control found to give quick and effective control of insect-pests. Though, high cost, environmental hazards, residual effects, unsafe for human and cattles etc. are some of the limitations in its use. It also requires technical knowledge to use and minor mistake results in heavy loss of crop. In general the use of chemicals is economical only in high value crops but if severity is more than there is no option other than chemical control.

The important insect-pests, their scientific name, nature of damage and control measures are as under:

Termites (*Odontotermes obesus* Ramb and *Microtermes obesi* Heomgr)

These are tiny grayish-white social insect lives in groups.

Nature of damage

Termites cause damage right from germination to subsequent growth stages of the crop. They cut the main root of the plant and ultimately damaged plant dry up. Such plants, if uprooted, show presence of termites and their galleries on damaged plants. Many times it causes severe damage, results in crop failure.

Control measures

1. In case of heavy infestation application of Methyl parathion dust (2%) or Endosulfan dust (4%) @ 25 kg/ha in soil at the time of field preparation.
2. Seed treatment with Chloropyriphos @ 10 ml or Endosulfan 15 ml/kg seed.

Cutworm (*Agrotis ipsilon* Ratt.)

Cutworm is common in whole country but serious in northern India particularly flood affected areas and soils rich in organic matter. Adult moths are grayish-brown in colour. The forewings are dirty brown. The hind wings are semi-hyaline white.

Nature of damage

The larvae are polyphagous feeders infesting wide range of hosts. The moths hide during day times and come out of their hiding places at night. The caterpillar cut the plant at base, below or just above the soil surface or may even cut the branches. The characteristic feature of damage is that the caterpillar drags the cut parts in the soil for feeding.

Control measures

1. Application of Methyl parathion dust (2%) @ 25 kg/ha in soil at the time of field preparation.
2. Seed treatment with chloropyriphos 20 EC @ 12 ml/ kg of seed.

Gram semi looper (*Autographa nigrisigna* and *Plusia orichalcea*)

Description

Though, Gram semi looper is a minor pest but it is widely distributed. Caterpillars cause the damage. The larva forms a loop during movement.

Nature of damage

The larvae feed on leaves, buds and flowers in the beginning and later on tender pods too. It eats up the whole pod including the seed leaving only a small part of pod attached to pedicle.

Control measures

1. Spray of 0.1 per cent Endosulfan (Thiodan 35 EC) is quite effective.

Aphids (*Aphis craccivora* Koch)

Description

Aphids occur throughout the country but usually in late sown crop. These are shining dark brown, black or greenish black pests.

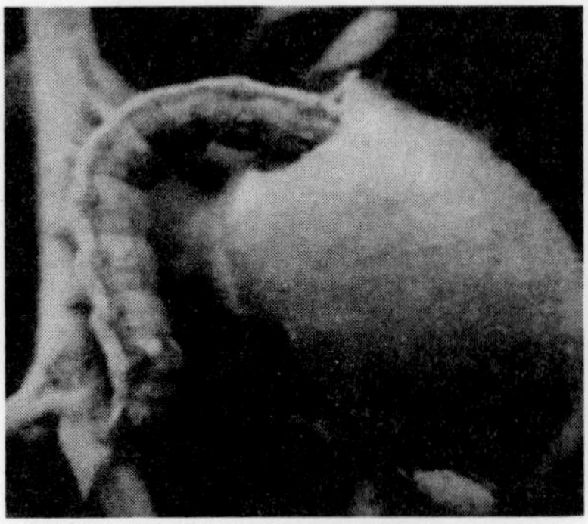

Pod borer

Cut worm

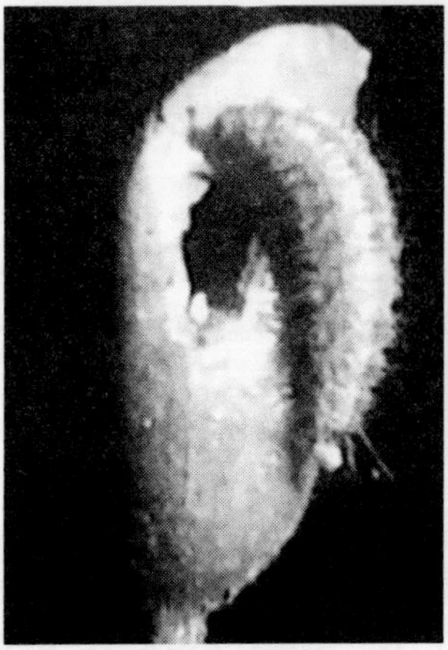

Plum moth

Bruchid damaged chickpea

Termite

Nature of damage

The nymphs and adults infest the leaves, tender shoots and pods. They cause damage by de-sapping of plants results into drying up of the tender shoots. In late stages the tender pods may shrival and dry seeds remain poorly developed.

Control measures

1. Spray of 0.1 per cent Dimethoate (30 EC) or Monocrotophos (36 WSC) and repeat at an interval of 15 days if needed.

Gram pod borer(*Helicoverpa armigera* Hubner)

Gram pod borer is cosmopolitan and is widely distributed in India.

Description

The moth is stoutly built and is yellowish brown in colour. The fore wings are marked with grayish wavy lines, black spots of varying size on the upper side, black kidney shaped mark and round spot on the underside. The hind wings are whitish and lighter in colour. The newly hatched larvae are sluggish and whitish-green in colour, become active after 3-4 hours. The full-grown larvae are 3.5-4 mm in length with pale green colour. Dorsal surface bears dark broken stripe, head reddish brown and legs are brown.

Nature of damage

The caterpillar is the damaging stage. Young larvae feed on tender leaves, buds, and flowers and subsequently bore into pods. Normally larvae remain hidden in the foliage of the crop unnoticed till pod formation. It causes the damage up to a great extent. A single larva may destroy 30-40 pods before it reaches maturity.

Control measures

1. Timely sowing *i.e.* up to mid October, growing of early cultivars complete podding by I week of March in northern region helps in escaping the peak attack of the pest.
2. Spray of 0.1 per cent Endosulfan (Thiodan 35 EC) or Monocrotophos (36 WSC) and repeat at an interval of 10 days if needed.
3. Dusting of Malathion dust (5%) or Endosulfan dust (4%) @ 25 kg/ha.
4. Mixed cropping with non-preferred host plants like barley, wheat, mustard and linseed.

REFERENCES

Chhabra, K.S.; Lal, S.; Kooner, B.S. and Verma, M.M. 1992. Insect pests of pulses. Identification and control manual. Punjab Agricultural University, Ludhiana and Directorate of Pulses Research, Kanpur.

Khedar, O.P.; Maloo, S.R. and Punia, S.S. 2012. Pulses in Rajasthan: An overview Pulses production Strategies. by Dr. P. Murugesa Bhoopati and Dr. M. Paramathma. Sri Shakthi Promotional Litho Process. S.F. No. 283 Masaniamman Nagar, Anna Nagar East, Edarpalayam, Coimbatore- 641025. **PP:** 191-226.

Khedar, O.P.; Singh, R.V.; Shrimali, Mahesh and Singh, N.P. 2008. Pulses: Status and Cultivation Technology. Aaviskar Publishers, Distributors, Jaipur (Raj.)

Kooner, B.S.; Taggar, Gaurav K. and Cheeme Harpreet K. 2007. Insect Pests of Pulses-Status and Management. Pulses at a glance. Pulses Section, Department of Plant Breeding, Genetics and Biotechnology, PAU, Ludhiana-141004.

Lal, S.S. and Ahmad, R. 2002. Integrated Insect Pest Management – Present Status and Future Strategies in Pulses. Proceedings of National Symposium on "Pulses for Sustainable Agriculture and Nutritional Security" held at New Delhi, April 17-19, 2001. **PP:** 101-110.

Rathore, Y.S. and Lal, Shankar 1994. Pest management through cultural practices in India. Indian Institute of Pulses Research Kanpur.

Srivastava, B.K. 1964. Pests of pulse crops. Entomology in India. **PP:** 83-87.

12

Integrated Nematode Management in Chickpea

S.M. Yadav, S.S. Shekhawat and O.P. Khedar

Plant parasitic nematodes are one of the major threat in chickpea cultivation. In light soils the problem of nematodes is much serious. The losses due to nematodes have been estimated from 5%-60%. The damage due to nematode is affected with the crop species, plant age, soil temperature, soil moisture, soil type and nematode population present in the soil. In different zones different nematode spp. causes economic damage to the chickpea (Table 12.1). The nematode species *Meloidogyne incognita, M. javanica and Heterodera cajani* cause damage in all the four zones (NWPZ, NEPZ, CZ and SZ) of the country.

NEMATODE – FUNGUS INTER-ACTION

Nematodes not only damage the crop but also interact with soil borne fungi and increase the severity of the disease. Some times when the fungus does not penetrate the roots of the plant in such case nematode penetrate the root first and then fungi penetrate the root easily and cause damage to the plant or in other words nematodes provide the path to the fungi to penetrate. Nematode interacts with different soil fungi and cause disease complexes in chickpea results heavy damage to the crop (Table 12.2)

Table 12.1. Major nematodes of chickpea in different zones in the country.

S.N.	Zone	Chickpea
1.	NWPZ	*Meloidogyne incognita, M. javanica, Tylenchorhynchus* spp., *Pratylenchus* spp., and *Heterodera swarupi* (in Rajasthan only)
2.	NEPZ	*Meloidogyne incognita* and *M. javanica, Tylenchorhynchus* spp and *Rotylenchulus reniformis*
3.	CZ	*Meloidogyne incognita, M. javanica* and *Pratylenchus* spp.
4.	SZ	*Meloidogyne incognita* and *M. javanica*

Table 12.2. Nematode-fungus inter-action in chickpea.

S.No.	Crop	Nematode	Fungus	Disease
1.	Chickpea	*Meloidogyne incognita*	*Fusarium oxysporum* f. sp. *ciceri*	Wilt
2.		*M. javanica*	*Fusarium oxysporum* f. sp. *ciceri*	Wilt
3.			*F. solani*	Black root-rot
4.			*R. bataticola*	*Dry root-rot*
5.			*Sclerotium rolfsii*	Collar rot

SYMPTOMS

Stunting, uneven poor plant growth in patches, pale yellowish appearance of plants, poor podding, reduced nodulation and reduced yield are some of the common characteristics of root-knot nematode attack. Root-knot nematodes characterized by the presence of simple multiple knots on the root system and can be seen by naked eyes. The size of knots varies depending on genotype reaction and nematode species. Roots swell due to nematode galls. Roots become black and brittle, results plant death in later stages.

Root-knot nematodes have also been found to reduce the number of nodules.

MANAGEMENT OF NEMATODES

A variety of nematode management approaches based on cultural, physical, chemical and biological methods have been found effective in reducing nematode population and improving crop production. It is quite impossible to eradicate nematodes completely except on very smaller area. Nematode population less than 2 larvae/g of soil does not affect much to the plant roots, but if the nematode population exceeds more than 2 larvae/g of soil, it causes damage to the plant and affects the yield, so we have to manage the nematode population below the economic threshold level (less than 2 larvae/g of soil). If nematode population exceeds too high, the death of plant will occur before maturity. None of the single control measure found to be effective and economically viable to control the nematodes population below economic tolerance level. Therefore, reliance has to be placed on a judicious combination of different practices suitable for location specific conditions. For nematode management we have to adopt following control measures:

A. Physical management

Solarization: For a smaller area this method may be used effectively. Solarization by covering the soil with transparent polyethylene sheets during the summers (April, May and June) significantly reduces the nematode population. Irrigation before solarization enhances the effectiveness of solarization.

B. Agronomic management

Agronomic management are an effective tool in controlling nematode population below economic thresh hold. The efficacy of cultural practices depends on soil type, nematode population, nematode species and agro-climatic condition of the soil. Summer ploughing, crop rotation and inter-cropping are some of the important practices by which nematode population can be managed easily. Before adopting any cultural practice it is essential to know the biology of nematode, host range, population density of nematode, soil type and agro-climatic condition.

a. **Summer fellowing and ploughing**: Summer fellowing and 2-3 deep ploughing helps in reducing root knot nematode population up to desired level and improves plant growth and yield too. Summer ploughing after irrigation is more effective than dry condition.

b. **Crop rotation:** Most of the nematodes like root-knot, cyst, reniform and lesion nematodes have wide but specific range of hosts so try to accommodate commercially acceptable and agronomically suitable non-host or poor host crop in crop rotation. Chickpea may be cultivated in rotation with cereals, millets oil seeds or onion (pearl millet/wheat/jowar/ maize/paddy/til/onion-legumes). Two-year rotation against cyst nematode and one-year rotation against root-knot and reniform nematode is sufficient, even then, long term crop rotation is more effective in controlling nematode infestation.

c. **Intercropping**: Inter-cropping is again an effective method of nematode control, because through inter-cropping we can incorporate some of the non-host crops, which helps in reducing the population of nematodes. Some of crop species like tagetes, sesame, mustard etc. showed antagonistic effect to the root knot and reniform nematodes. Inter-cropping of legumes with above crops helps in reducing nematode infestation, however, appropriate crop geometry is essential to avoid shading effect. Inter-cropping of mustard with chickpea and pea helps in reducing the population of root-knot nematodes.

d. **Cropping system**: Cropping system also plays an important role in nematode management. If host crops (Vegetables) are grown prior to chickpea results in heavy infestation of nematodes. In such cases rapid build-up of root-knot nematode occurs. If non-host crops (pearl millet, wheat, jowar) are grown prior to chickpea, it helps in reducing the nematode population. By including cereal or grasses in the cropping system root-knot nematode in chickpea is effectively controlled. Sesame, mustard and winter cereals are poor hosts of *M. javanica* and *M. incognita* and if included in the system reduce the nematode population in 2-3 years.

e. **Avoid mono-cropping:** Mono-cropping helps in build-up of nematode population so avoid mono-cropping.

f. **Weed free cultivation**: Some of the weeds acts as host plant for some of the nematodes so weed free cultivation helps in drastic reduction of nematode population because some of the weeds act as host for nematode. Clean cultivation helps in nematode management up to a remarkable level.

g. **Organic amendments:** Application of organic manure helps in reducing plant parasitic nematode up to a satisfactory level. Application of non-edible deoiled seed cakes of neem (*Azadirachta indica*) karanj (*Pongamia glabra*), mahua (*Madhuca indica*), castor (*Ricinus communis*) etc. improves soil fertility and reduces nematode population through their decomposition products promoting the antagonistic microbial activity and are toxic to nematodes. The nematode population reduces significantly if oil cake of neem, caster, mustard or duan (Eruca sative) are incorporated in the soil. *Neem* cake @ 500 kg/ha at the time of field preparation.

C. Biological management

Bio-control agent has become important now a day. The micro-organisms like *Pasteuria penetrans, Paecilomyces lilacinus, Verticilium chamydosporium, Trichoderma viride, T. harzianum, Aspergillus niger* etc. has great promise against nematode management. However, prior testing of such organisms to human and cattle health is must. Application of *Pseudomonas fluorescens* + *Trichoderma viride* @ 2.5 kg/ha is also effective in controlling nematodes.

D. Management through plant-host resistance

Resistant varieties are an effective and cheapest approach for nematode management effectively and also prevents nematode reproduction. Resistance varieties (like RSG 888) are eco-friendly and may be grown successfully on nematode infested lands. It does not require any special technical guidance and equipment for its use. However, initially perfect screening and testing is must for the development of resistant material. Though, there are so many genotypes available in legumes but even then it is not up to mark till date and needs more perfection.

E. Chemical management

Like other biotic stresses in nematode also chemical control found to give quick and effective control. Though, high cost, environmental hazards, residual effects, unsafe for human and cattle etc. are some of the limitations in its use. It also requires technical knowledge to use and minor mistake results in heavy loss of crop. The use of chemicals is economical only in high value crops, whereas, in case of chickpea in general it is not cost effective but if severity is more there is no option other than chemical control. Spot or row application of chemical is more effective than application in the whole field. In case of severe infection soil application of Carbofuran 3G @ 45 kg/ha (1.5 kg ai/ha) at the time of field preparation is recommended.

F. Seed treatment

Seed treatment is again cheaper and effective way of nematode management. It provides initial protection to the young plants. For effective nematode management following measures may be adopted:

i. Seed treatment with Carbosulfan 25 ST @ 3% ww (30 g/kg of seed).

ii. Seed treatment with Carbofuran 3G @ 2%.

G. New approaches

With the development of technical knowledge there are great possibilities to develop bio-control agents or bio-rational novel nematicides may target the nematode plant inter-action and bio-engineering for nematode resistance. Genetic transformation may also prove to be mile stone in nematode management in the coming time. Techniques of molecular biology may help to in the identification of plant-parasitic nematodes.

REFERENCES

Ali, S.S and Gaur, H.S. 2007. Management of plant parasitic nematodes in pulses. Souvenir. National Symposium on Legumes for Ecological Sustainability: Emerging Challenges and Opportunities held at IIPR, Kanpur, Nov. 3-5, 2007 **PP:** 24-29.

Ali, S.S. and Gurha, S.N. 1994. Interaction between *Fusarium oxysporum f. sp.ciceri and Meloidogyne javanica*. Proceedings International Symposium on Pulses Research, April 2-4, 1994, New Delhi.

Ali, S.S. and Singh, Bansa 2007. An overview of nematode constraints in pulse and their management. Journal of Food Legumes. 20(1): 1-8.

Bhatnagar, C.P.; Handa, D.K. and Mishra, A. 1988. Kabuli chick-pea mutants resistant to root knot nematode, *Meloidogyne javanica* International Chickpea Newsletter 19: pp 16-17.

Gaur, H.S.; Ali, S.S. and Haseeb Akhter 2003. Integrated Nematode Management in Pulses. Pulses in new perspectives (eds. by M. Ali and Shiv Kumar) ISPRD, Indian Institute of Pulses Research, Kanpur. **PP:** 457-470.

Khedar, O.P.; Maloo, S.R. and Punia, S.S. 2012. Pulses in Rajasthan: An overview Pulses production Strategies. by Dr. P. Murugesa Bhoopati and Dr. M. Paramathma. Sri Shakthi Promotional Litho Process. S.F. No. 283 Masaniamman Nagar, Anna Nagar East, Edarpalayam, Coimbatore- 641025. **PP:** 191-226.

Khedar, O.P.; Singh, R.V.; Shrimali, Mahesh and Singh, N.P. 2008. Pulses: Status and Cultivation Technology. Aaviskar Publishers, Distributors, Jaipur (Raj.).

Sharma, S.B. and Nene, Y.L. 1990. Effect of soil solarization on nematodes parasitic to chickpea and pigeonpea. *Journal of Nematoleranceogy*. 22(45) 644-658.

13

Seed Production in Chickpea

S.S. Shekhawat, O.P. Khedar, S.S. Punia, Yogendra Singh, R.V. Singh and Hoshiyar Singh

Seed is the backbone of sustainable crop production. It is a fertilized and ripened ovule that contains a viable embryo and is able to germinate and produce a plant. It is initial and the basic input responsible for improving productivity and quality of produce. The quality seed is the cheapest input with maximum return and may increase the productivity up to great extent. Therefore, emphasis should be given on the quality seed production of pulses because these are poor man's crops and quality seed may change the economic status of the farmers. The investment on the quality seed is very much important in crop production. The need of quality seed having desired characteristics is well recognized for increasing the crop yield in any agro-climatic region. It is also a trigger point which sets in motion the process of technological change. Seed account for a small part of the total cultivation expenditure. Without good quality seed, the investment on fertilizer, irrigation, pesticides and other inputs will go in vain. Neither the poor quality seed of superior variety nor the high quality seed of inferior variety can serve the farmer satisfactorily. Thus, the basic objective of any seed production programme is timely supply of high quality seed of superior variety at remunerative price and at required place. Under the traditional system of farming most of the farmers used to retain a part of their produce for use as seed. Advances in research resulting in the evolution of high yielding, short duration, dwarf, drought tolerant, and disease and insect-pest resistant varieties have added a new dimension to the pulse production. The supply of improved varieties is important not only for realizing full potential of varieties but for retai-ning faith of farmers in high yielding technology. The seed should be genetically pure and free from seed borne diseases.

Recognizing the importance of quality seeds in agriculture, the National Seeds Corporation (NSC) was set up in 1963 for production and distribution of quality seeds. Later, States Farms Corporation of India (SFCI) was set up in 1969 to develop modern farms, mainly for production of quality seeds. In addition, States Seeds Corporations (SSCs) have been set up in different states. Now a day apart from above institutions ICAR institutions, State Universities, and Private seed producers are engaged in the production of quality seed.

SEED QUALITY

The importance of quality seed was recognized with the beginning of human civilization or we may say with the cop production started. In general farmers select the best plant for planting of next year. In early age, bold seeded and good looking plants were the only criteria for quality seed but with the development of seed science the parameters of quality seed have been changed. Now-a-days quality seed includes genetic and physical purity, viability, germination, and moisture and seed health. Genetic purity means it should be true to type and there should not be off-types or other variety seed. Germination percent is an important criteria and it should not be less than the prescribed criteria. Moisture content affects the seed quality up to great extent because if moisture content is not according to prescribed criteria it will invite seed borne diseases and ultimately it will deteriorate the germination and viability of the seed. Features of quality seed are as under:

1. It must be genetically true to type and full-fill general identity wart. Distinctness, uniformity and stability (DUS).
2. It must be viable and germination should be as per standards.
3. It must be physically pure.
4. It must be free from grain pests.
5. It must be free from seed-borne diseases, objectionable weed seed and other crop seeds.
6. It must meet out the germination and purity standards

CATEGORIES OF SEED

At the time of release of a variety only nucleus seed is available with the breeder. The commercial seed is produced after a series multiplication stages. The different stages or categories involved in seed production chain are:

A. Nucleus/basic seed

This is I category in seed multiplication chain produced in close watch of breeder by the institute. For basic seed production a number of individual plants (400-500) are to be selected from the existing nucleus seed plot. These plants are harvested and threshed individually. This is called Nucleus Seed Stage I commonly referred as NSS I. Differences in early plant growth habit, leaf colour, later growth habit, plant height, days to flowering, days to maturity, disease incidence are to be looked for and the rows containing off type plant or plants are rejected as and when noticed. Rest of the rows conforming to the variety are harvested and threshed individually. Produce of plant progeny rows is also screened for seed characters. The seed lots retained after the screening are bulked to constitute the nucleus seed of the variety and used for breeder seed production. Sometimes when seed requirement is more then we go for two stages NSS I (Nucleus Seed Stage I) and NSS II (Nucleus Seed Stage II). In NSS I individual line bulk is harvested and kept separately and used for NSS II production in next year. In the next year individual line bulk from NSS I are planted in smaller plots and if any plot shows any variation the whole plot is rejected and remaining plots are harvested, bulked and threshed. This is the seed of NSS II and may be used for production of breeder seed. This category of seed is not in the preview of seed law enforcement practically. Government has not prescribed any criteria but supposed to be 100% pure (genetically and physically) because it is produced under the supervision of concerned breeder/institute.

B. Breeder seed

This is II category in seed multiplication chain and is the progeny of nucleus seed. It is the mandate of Indian Council of Agricultural Research (ICAR) and is being undertaken with the help of ICAR Research Institutions, State Agricultural Universities (SAUs), Sponsored breeders recognized by selected State Seed Corporations and Non-Governmental Organizations (NGOs). It is produced under the direct supervision of qualified breeder. It must be genetically 100% pure. The sowing should be done in such a way that it facilitates the roguing and row spacing should be sufficient to permit examination of off-types. Breeder seed is the most important step in the seed multiplication chain. Breeder seed crop is monitored by a joint monitoring team of breeder of the variety, the concerned Project Director or his nominee, representative of NSC & State Seed Certification Agency. Four to minimum of two monitoring at the time of flowering and at maturity are must. Breeder seed should be supplied with Breeder's golden yellow tag as per the guideline given in Indian Minimum Seed Certification Standards.

C. Foundation seed

This is progeny of breeder seed and is III category in seed multiplication chain. The responsibility for production of foundation seed has been entrusted to the NSC, SFCI, State Seeds Corporation, State Departments of Agriculture and private seed producers, who have the necessary infra-structure facilities under technical control of plant breeders are technical officers. The seed production plots are dully certified by the concerned Seed Certification Agencies. Foundation seed is required to meet the standards of seed certification prescribed in the Indian Minimum Seeds Certification Standards, both at the field and laboratory level. Four to minimum of two monitoring are must. This produce is the source for production of certified seed.

D. Certified seed

This is IV and last category in seed multiplication chain and progeny of foundation seed. It is for distribution to the farmers. It must meet the standards of seed certification prescribed in the Indian Minimum Seeds Certification Standards, 1988. The seed of this category is normally produced by State and National Seeds Corporation and private seed companies on their own farms or farms of progressive cultivators. This seed is used for commercial production or it is the seed which is actually sold to the farmers. The containers shall carry white tags. The production and distribution of quality/certified seeds is primarily the responsibility of the State Governments. Certified seed production is organized through State Seed Corporation, Departmental Agricultural Farms and Cooperatives etc.

Truthfully labeled seed (TL Seed)

This category of seed is not mentioned in seed act but ICAR institutes, SAU's, Public and private companies are selling their seed under this category. This category of seed can be produced from any class of seed may be from nucleus, breeder, foundation or certified. No certification is required in TL seed but labeling is compulsory. It must meet out the seed standard as prescribed for certified seed. In general concerned breeder or institute produces such type of seed under their direct supervision so this category of seed is also reliable.

MAINTAINANCE OF GENETIC PURITY

Genetic purity is most important for the maintenance of a variety. Following points should be taken in care during seed production programme:

1. Adequate isolation distance to prevent contamination.
2. Rouging of seed fields at proper stage.
3. Periodic testing of varieties for genetic purity.
4. Avoiding genetic shifts by growing crops in areas of their adaptations only.
5. Certification of seed crop to maintain genetic purity and seed quality.
6. Adopting the generation system.

PRINCIPALS OF QUALITY SEED PRODUCTION

Seed is the initial and most important input in crop production and the use of quality seeds may alone increases productivity to the extent of 15-20 per cent. The seed quality is very important in agriculture and it should not be compromised at any cost. Without quality seed there is no use to waste other inputs. Following principles should be take care during quality seed production:

A. Genetic principals

Chickpea is self-pollinated crop. Though, cross pollination also take place up to some extent, mostly caused by insects. So population of pollinating agents and prevailing environment also affect degree of cross pollination. The genetic purity of the seed is affected by so many factors like:

a. Seed standards

The seeds meeting required standards of purity (genetic and physical purity), germination and seed health. Quality seed production should meet out the following standards:

1. Genetic purity

Seed must be genetically true to type and expected to possess certain diagnostic characters to distinguish from other varieties. It should fulfill general identity i.e. distinctness, uniformity and stability (DUS).

2. Germination and vigor

It must be viable and meet out the germination and purity standards.

3. Physical purity

Physical purity means there should not be any kind of mixture like other crop seed, weed seed and inert matter. It must be free from objectionable weed seed and other crop seeds.

4. Seed health

The healthy seed not only helps in raising disease free crop with good germination, but also helps in checking further spread of disease.

b. Seed and its source

The source of seed is very important in quality seed production of any crop. It should be pure and obtained from approved source. It should meet out the standards of quality seed. Following points should be kept in mind at the time of purchase of seed.

1. Should be of appropriate class.
2. Tags, labels and seals of seed bags should be intact.

3. Validity period should not be expired.
4. Whole seed should belong to same lot.

c. Prevention from mechanical mixing

Mechanical mixing is very important in seed production programme because all seed production efforts will render futile if sufficient care is not taken in field operations to avoid mechanical mixing. The machines used in harvesting, threshing, transport and processing should be clean thoroughly. The field operations which require seed producer's attention in this regard are sowing, rouging, harvesting, threshing, processing and bagging.

d. Isolation

Though, chickpea is self-pollinated crop even then, to avoid mixture from adjoining plots proper isolation distance should be maintained. Isolation requirement for chickpea is given in Table 13.2. Sources of contamination could be other varieties of the same crop, same variety not conforming to varietal purity requirements of the category of seed under production or weeds.

e. Roguing

Roguing is the process of removing undesirable plants from the seed crop to maintain genetic purity of a variety at field level. It should be a regular process to maintain genetic and physical purity. The rogues may be weeds, off-types, other variety plants; plants affected with diseases, pollen shadders etc. Any plant which does not conform to the characteristics of the variety is called an off-type. Weeds, abnormal, volunteer, diseased and other species or variety plants should be removed up to maturity. Roguing should be done at minimum possible interval to maintain genetic and physical purity. The crop should be harvested at proper stage of maturity. As a general rule, the off-types should be removed and taken away from the seed production plot and destroyed. Roguing should be done in the morning hours and the back of the person doing roguing should be towards the sun. This facilitates easier detection of off-types. Following characters should be observed during roguing of chickpea crop (Table 13.1).

Table 13.1 Characters to be observed during roguing at different stages.

S.N.	Stages	Characters to be observed
1.	Seedling	Pigmentation on stem, angle of cotyledon, leaves hairiness, leaf colour.
2.	Vegetative	Plant colour, branching pattern, pubescence on stem, leaf colour, presence of pigmentation.
3.	Flowering	Type of inflorescence, flower position, flower colour, plant height, weeds, infected plants.
4.	Maturity	Pod shape, pod size, pod colour, shattering habit, plant height, maturity habit, foliage colour at maturity.
5.	Post harvest handling	Pod characters, seed size and shape, seed shining or dull.

f. Seed Certification Standards

Seed certification standards helps in maintaining genetic and physical purity. There are two types of standards: field standards, which are applicable to standing crop and seed standards applicable at seed level. Field standards include isolation requirement, maximum permissible level of off-types (Table 13.2), inseparable other crop plants, objectionable weed plants, plants infected by seed borne diseases etc. Seed standards include genetic purity, physical purity, germination, other crop seeds, weed seeds, moisture content etc (Table 13.3). In seed production of any category, seed certification standards should not be compromised and it should be followed strictly.

Table 13.2. Isolation distance and off-type plant limit in foundation and certified seed production plots of chickpea.

S. No.	Crop	Foundation seed		Certified seed	
		Minimum isolation distance (m)	Off-type plants (%)	Minimum isolation distance (m)	Off-type plants (%)
1.	Chickpea	10	0.1	5	0.2

Table 13.3. Seed standards of chickpea.

S. N.	Crop	Type of seed	Purity (%)	Other crop seeds per kg	Weed seed per kg	Other variety seed per kg	Inert matter (%)	Germination (%)	Moisture (%) (Normal packing)	Moisture (%) (Vapour-proof containers)
			Min.	Max.	Max.	Max.	Max.	Min.	Max.	Max.
1.	Chickpea	F	98	0	0	5	2	85	9	8
		C	98	5	0	10	2	85	9	8

Here: F- Foundation and C- Certified

g. Seed testing

Seed testing is an essential and indispensable part of seed certifi-cation and quality control. After processing of a given lot, the representative sample is sent to notified seed testing laboratory for its analysis. Seed testing is done as per ISTA Rules. ISTA 1993 has developed uniform rules and regulations for testing. If the seed meets the prescribed seed standards then a given seed lot is approved and tags (of prescribed colour and size) and certificates are issued by the certification agency to the seed producer. Such type of seed is ready for the sale to the cultivators. The validity period is nine months from the date of test at the time of initial certification. The validity period can be further extended for six months provided on retesting the seed conforms to the prescribed standards. Improper handling of seeds also causes mechanical injury and lower down its germination and storability. Therefore, during post-harvest processing and storage, seed must be handled properly and protected from high relative humidity and temperature, insect pests and rodents. The main objective of seed testing is to serve the producer, seed industry and the cultivator both.

h. Labeling of seed

Labeling is must before supply of seed or in other words during seed supply it should carry prescribed label on the bags. The label is self explanatory about the information as per certification standards like name of crop, category of seed, variety, purity (%), other crop seeds per kg, weed seed per kg, other variety seed per kg, inert matter (%), moisture (%) in normal packing, germination (%) etc. (Table 13.3). Label colour and size have also been specified for each category of seed. If the tag is to be fixed on a smaller container then the size of the tag can be reduced proportionately (except for breeder seed). However, length and breadth ratio and contents would remain the same. Label colour and size should be as per Table 13.4.

Table 13.4. Label colour and size of label for different categories of seed.

S. No.	Class of seed	Label Colour	Label Size
1.	Breeder	Golden Yellow	12 cm x 6 cm
2.	Foundation	Both side White	15 cm x 7.5 cm
3.	Certified	Both side Azura Blue	15 cm x 7.5 cm
4.	T.L. Seed	Opel Green	15 cm x 10 cm

i. Grow-out test

Grow-out test is a pre-requisite for grant of certificate. It is must to determine the genetic purity of a given seed lot. The seed sample is subjected to grow-out test to check genetic purity before it is finally approved as certified seed. The samples for grow-out test are to be drawn simultaneously with the samples for other quality tests. Grow-out test is done in the off-season by some State Seed Certification Agency. If any seed sample fails in grow-out test, the seed is rejected and it is not fit for further crop production. The samples passed in grow-out test are approved for its use as seed. The specifications for some of the crops are as under (Table 13.5).

The seed rate may be adjusted depending on germination percentage of individual sample and may be done by dibbling. Thinning is not recommended. The test crop may be raised along with control.

Table 13.5. Field specifications for grow-out test in chickpea.

S. No.	Crop	Row length (m)	Plant to plant distance (cm)	Row to row distance (cm)	Plot to plot distance (cm)	No. of rep.
1.	Chickpea	6	10	30	60	2

j. Field inspection

Field inspection ensures genetic and physical purity of the seed at field level. Field inspections are required to be carried out in the recommended/prescribed crop stages. All the factors in the field may not be apparent or may not be capable of affecting seed quality at a particular stage of crop growth. The numbers of inspections given in the Minimum Seed Certification Standards are mandatory. The details of crop stages and factors to be observed in crop stages are as below:

1. Pre flowering stage (Vegetative stage)
2. Flowering stage
3. Post flowering stage
4. Pre harvest stage

5. Harvest stage

B. Agronomic principles

a. Selection of seed plot

The seed plot must fulfill the following conditions.

1. Soil should be clean, friable, leveled and well drained.
2. The same crop should not have been grown in the previous season.
3. It should be free from soil borne diseases, nematodes, insects-pests and objectionable weeds.
4. Soil should be free from volunteer plants.
5. It should be feasible to isolate the plot form any contaminant.
6. Soil should have good water retention capacity and proper drainage facility in case of heavy rains.

b. Field preparation

The field should be weed free and well leveled before sowing. For preparation of field ploughing and harrowing should be done to make the field friable.

c. Seed treatment

Seed treatment with fungicide protect the crop from seed borne diseases, insecticide from insect-pests and nematicide from nematodes, whereas, *Rhizobium* culture enhances the nitrogen fixation through biological nitrogen fixation, PSB converts insoluble phosphates into soluble form and make it available to the plants and PGPR promotes the plant growth, which consequently increases the yield. It also helps in improving the quality of seed.

d. Time and method of sowing

In general seed crop should be sown one week before normal crop to avoid insect-pest and disease attack. It also helps isolating from the normal crop up to some extent. Seed crop should be sown in rows only to conduct field inspection and rouging properly. Row sowing also helps in adopting plant protection measures properly. Sufficient moisture should be ensured before sowing to get good germination.

e. Fertilization

There are 16 essential elements required for plant nutrition. Prescribed dose of nutrients should be added to the soil for good yield. Generally micronutrients are present in the soil in sufficient amount but its deficiency results in non-availability or imbalance of other nutrients also which affects the seed production. The application of micronutrient helps in seed filling, seed yield, oil content, protein content, seed weight, seed density, seed germination and seed vigour.

f. Irrigation

Number and interval of irrigation depends on soil type and weather condition. It is very much essential to irrigate the seed crop time to time during growth stages for good production of seed. Irrigation at the time of germination, flowering, seed filling and maturity increases the yield up to great extent.

g. Weed control

The weed plants have wider adaptability over cultivated crops. They can survive in adverse condition even where crop plants cannot perform their normal life. They also have high growth rate as compared to cultivated crops. They compete for space, nutrient, moisture, light etc. and ultimately reduce the seed yield. Therefore, it is very much essential to keep the seed crop weed free throughout the growth period. Due to high value of seed, weed control has greater significance in seed production plots. Weeds in early stage damages the crop up to great extent so special care should be taken to keep the crop weed free up to 40-45 days. To keep the crop weed free at least two hand weeding or pre-emergence application of weedicide accompanied with hand weeding is must.

h. Harvesting and threshing

Though, timely harvesting is an important operation in any crop production programme but in case of seed production it is essential to prevent the crop from bird damage, damage from untimely rains, storage fungi, and store grain-pests and to improve seed quality. After harvesting the crop should be threshed, winnowed, cleaned and dried well before processing.

C. Plant protection principles

The seed crop and seed should be insect-pest and disease free start from sowing up to seed distribution. It results in good production, better seed quality, good germination and vigorous crop.

a. Disease, nematode and insect-pest control

The plant protection measures should be adopted well in time because delayed adoption of plant protection measure does not help in controlling diseases, nematodes and insect-pests and results in low production of poor quality seed. In case of application of fungicide, nematicide or pesticide the dose should be perfect because over dose will adversely affect the crop and produce, whereas, lower dose will not be able to control the biotic stresses effectively so judicious dose should be used for plant protection.

b. Pre-harvest sanitation

Pulses suffer more by store grain pests and they infest the seed at the field level. To avoid bruchid infestation in the store following protection measures should be adopted in the standing crop.

1. Spray of Melathion (0.07%) either singly or in combination with Carbandazim (0.1%) two times at an interval of 7 days before harvest of the crop.
2. Spray of Trizophos @ 200 ml/ha or soluble sulphur @ 0.3%.
3. Spray of *Nimboli* extract @ 5%.

It prevents the seed from the attack of biotic agents for a longer time and also increases the number of pods and yield.

D. Post harvest principles

a. Seed processing, packaging and storage

Seed processing is an important process in seed production, it is an integral part of seed industry and responsible for upgrading the physical quality of seed. It is mainly carried out for removal of foreign materials, chaff, stem pieces, leaf parts; soil particles weed seed, broken seed etc. and helps in improving seed quality, viability and storability of seed. The objectives are to improve the physical quantity of seed. The seed crop as it comes from the field is never pure. The harvested seed contains immature seeds, seeds of other crops, broken seeds, weed seeds, soil particles and other inert materials. These contaminants must be removed after harvesting and threshing to get pure crop seeds for replanting. The main purpose of processing is to maximize germination, seed recovery and physical purity in addition to minimize seed loss, mechanical damage and insect infestation. The main objectives of processing are:

1. To remove foreign materials such as broken straw, stones, dust and other plant seeds.
2. To remove weed seed and broken seed.
3. To obtain optimum seed size.

4. To remove excess moisture for better storability and quality of seed.

Packing is also important for the prevention of adulteration with the inferior seeds. The seed should be pack in reduced moisture content to safe limits in vapour proof containers. A label is to put on each tested seed packet by the seed certification agency. After packing seed should be stored in ventilated, damp proof and insect free storage until it use in the coming season. The precautions suggested in safe storage should be adopted while storing the seed. The time and money spent on seed production go waste if the harvested seed is not properly handled.

b. Post harvest handling of seed

1. Field should be free from weeds.
2. The transportation vehicle should be clean properly.
3. Threshing floor should be clean.
4. Before threshing another variety, threshing floor and transportation vehicle should be clean again properly.
5. The thresher should be clean thoroughly.
6. At the transportation of seed from the threshing floor to the seed store more than one variety should not be transported at a time.
7. At the time of processing of seed the floor and machines should be clean properly.
8. On packing only new bags should be used to maintain the purity of the seed.
9. The label should be put on the bags properly.

E. Economic principles

Economic aspect is again important in seed production programme. The seed production programme becomes economically viable only when benefit-cost ratio is high. It is greatly influenced by seed multiplication ratio, input cost, managerial cost; fixed cost, seed demand and market price. Before adopting seed production programme whole programme should be chocked and it should be done in the supervision of experts only.

SEED CERTIFICATION

Seed certification is a scientific and systematic procedure for genetically pure seed. Seed certification is not a seed programme in itself. For seed industry, seed certification forms a basis for quality differentiation, particularly in the early stages of a programme. In our country seed certification is linked with notification of a variety.

A. Principles of seed certification

The fundamental principles involved in this process are:

a. Secure basic seed from a recognized source.

b. Maintain genetic identity of a variety with respect to DUS.

c. High degree of genetic purity.

d. High degree of germination.

e. Free from any contamination like mechanical/ genetic.

f. Free from seed borne diseases.

g. Seed produced must qualify the minimum standards under field and laboratory conditions.

h. To make available quality seed to farmers.

For a seed certification programme to be successful it is essential to have:

a. Sound crop improvement programme for regular flow of stable, superior varieties.

b. Well-qualified and trained staff to develop sound seed certification standards and procedures.

c. Infra structure for efficient management of post harvest processing of seed crop.
d. To have trained seed producers, arranging financial resources, marketing system, enforcement responsibilities and constant liaison with related agencies.

B. Phases of Seed Certification

a. Receipt and scrutiny of application including eligibility of the variety, verification and establishment of seed source and other requirements.
b. Field inspections to verify conformity to the prescribed field standards.
c. Supervision at post-harvest stages including processing and packing.
d. Seed sampling and analysis including genetic purity test and/or seed health test, if any, in order to verify conformity to the prescribed standards
e. Grant of certificate and certification tags, tagging and sealing.

SEED PRODUCTION NETWORK IN INDIA

Sufficient infrastructure facilities were developed with the support of ICAR institutions, SAU's, NSC, SFCI, SSC and Private seed growers. The infrastructure includes development and creation of seed production farms, irrigation facilities, seed planter, harvester and threshers, storages, processing equipments and machinery and equipments of seed testing laboratories. The BSP and STR units were established at SAU's and ICAR institutes. Now in India we have largest seed production network of the world engaged in quality seed production of important crop varieties. Almost all the states have established the State Seed Corporations for production and marketing of seed and State Seed Certification Agencies for quality control through certification. Besides the public sector private sector also play an important role in seed production and marketing, though, it is mainly concentrated on hybrid seed production and least involved in the seed production of self pollinated crops including legumes.

REFERENCES

Atwal, S.S. 2004. Production of quality seeds in field crops: An overview. Lecture delivered during Winter School on "Development of Quality Seeds: Principles and Practices" (8-28 September, 2004). IARI Regional Station, Karnal, Haryana.

Khedar, O.P. and Maloo, S.R. 2012. Seed production in pulses. Pulses production Strategies. by Dr. P. Murugesa Bhoopati and Dr. M. Paramathma. Sri Shakthi Promotional Litho Process. S.F. No. 283 Masaniamman Nagar, Anna Nagar East, Edarpalayam, Coimbatore- 641025. **PP:** 347-362.

Khedar, O.P.; Singh, R.V.; Shrimali, Mahesh and Singh, N.P. 2008. Pulses: Status and Cultivation Technology. Aaviskar Publishers, Distributors, Jaipur (Raj.).

Kumar, D. and Singh, N.B. Manual for seed production of arid legumes. All India Co-ordinated Research Project on Arid Legumes, CAZRI, Jodhpur (Raj.).

Law of seeds. 2005. Asia Law House. Hyderabad.

Renuga Devi, J. and Vanangamudi, K. 2008. Production practices and maintenance of nucleus and breeder seed. Advances in seed science and technology by K. Vanangamudi, N. Natrajan, K. Natrajan, A. Bharthi, R. Umarani and T. Saravanan. Agrobios (India), Jodhpur. **PP:** 335-362.

Srimathi, P. and Natarajan, K. 2008. Steps in maintenance of genetic purity. Advances in seed science and technology by K. Vanangamudi, N. Natrajan, K. Natrajan, A. Bharthi, R. Umarani and T. Saravanan. Agrobios (India), Jodhpur. **PP:** 535-541.

Tanwar, N.S. and Singh, S.V. 1988. Indian Minimum Seed Certification standards. The Central Seed Certification Board, Department of Agriculture & Cooperation, Ministary of Agriculture, Government of India, New Delhi.

Vanangamudi, K. and Natarajan, K. 2008. Seed technology and Management - Indian situation. Advances in seed science and technology by K. Vanangamudi, N. Natrajan, K. Natrajan, A.

Bharthi, R. Umarani and T. Saravanan. Agrobios (India), Jodhpur. **PP:** 9-20.

Yadav, R.N. 2004. Quality seed production in pulses with special reference to green gram. Winter School on "Development of Quality Seed-Principles and Practices" from 8-28 Sept. 2004 at IARI Regional Station, Karnal, Haryana.

14

Safe Storage in Chickpea

S.S. Bareth, S.S. Shekhawat O.P. Khedar and R.S. Sain

The success of chickpea production largely depends on efficient storage, because if storage is not proper then there is no use of production. The purpose of seed storage is to maintain the seed in good physical and physiological condition from the time they are harvested until the time they are planted. The climatic conditions prevailing at the time of maturity of the crop, sharp changes in environment, diseases and insect-pest incidence determine the storability of the produce. Unfavorable weather conditions like rain, hail, storm, high humidity and time of maturity of crop influence the storability. The need of storage of seed after processing arise due to the fact that there is a gap between harvesting and planting of next crop or in other words seed storage is required to protect the seed up to nest season to kept it viable. The success of seed production depends on efficient storage. A number of abiotic (moisture content, temperature, relative humidity of atmosphere) and biotic (insect-pests, mites, fungi, rodents, birds) factors affect the quantity and quality of the grain greatly. A number of biotic agents, store grain pests and rats are the major threat to the storages. At the time of storage moisture content should not be more than 10 per cent. The main objective of seed storage is to maintain the germination capacity and purity of seed. Generally the processed seed is stored for the coming season hence good storage is a basic requirement for healthy and viable seed. Well-dried seeds should be filled in insect free, clean and dry gunny bags and then store in clean and dry place. Dry seed stored under low temperature and humidity live longer and maintain high quality. The unutilized seed should be carried over for the subsequent season to serves the purpose of seed security. An ideal storage should satisfy the following requirements:

1. It should provide maximum possible protection from ground moisture, rains, insect pests, diseases, rodents, birds, fire, etc.
2. It should provide the necessary facility for inspection, disinfection, loading, unloading, cleaning and reconditioning.
3. It should protect seed from excessive moisture and temperature favourable to both insect and mould development.
4. It should keep the grain viable and maintain its quality.
5. It should provide better storage from processing till sowing of next crop.
6. It should provide buffer stocking for seed security.
7. It should be economically viable and suitable for a particular situation.

Till date 70% of the grain is stored at farmer's level, whereas, in case of chickpea this percentage is much more. In general the seed stored at farmers level in the storages which are not scientific, hence losses are more. With the awareness about the importance of seed in crop production the demand of seed has increased greatly. Therefore, it is an urgent need to store the seed in scientific way to full-fill the increasing demand of seed in the coming years. For effective storage of seed following points should be kept in mind.

STORAGE BAGS/CONTAINERS

Normally grains are stored in bags made of jute, canvas or low density polythene sheet or in containers made of metal, wood or mud, however, bags are in general practice. Bulk storages are not in practice in India. Use of new bags helps to eliminate the possibility of mechanical mixing or insect infestation. Old or used bags should be cleaned properly and treated with suitable insecticide before use.

HYGIENE OF STORE

Thorough cleaning should be done before storage of seed to avoid infection of insect-pests and microorganisms. Spilled over seed should also be removed before storage of new seed.

STACKING OF SEED BAGS

The bags filled with grain should always be stacked on wooden pallet (platform). This prevents the seed from moisture through floor, facilitates cleaning and circulation of air and fumigant at the time of fumigation. Avoid use of hooks at the time of handling of bags. Alley space should be left between seed-stacks and walls of the store. Normally 50-60% area is utilized for storage of seed. It is also observed that seed, stored separately helps in better management of insect-pests because different crop seeds are attacked by different pests and needs different treatments.

FACTORS AFFECTING SEED STORAGE

a. Abiotic factors

The abiotic factors like temperature and grain moisture largely affects the seed storability. In general cool and dry storage conditions are safe for seed storage. The important abiotic factors help the insect-pest activity and deteriorate the grain quality are:

i. Temperature

At higher temperature (>20°C) most of the microbes, mites and insect-pests multiply and infect the stored grain or seed. The optimum range of temperature for most of the insect-pests is 28-32°C, whereas, temperature less than optimum range may decrease their feeding activity, prolong developmental period and may cause mortality due to starvation. Low temperature below 20°C is good for safe storage of seed. With the reduction of 5°C in temperature, the storability gets doubled.

ii. Humidity

High humidity coupled with high temperature deteriorates the seed quality quickly. With the reduction of 1% in humidity, storability gets doubled.

iii. Seed moisture

Seed moisture content directly affects the storability of the produce. The moisture requirement of different insects differ from each other, however, all the insects needs moisture more than 10% for their normal physiological activities. The seed moisture content 12-14% is congenial for the development of insects. Seed moisture content below 10% is considered good for storability. The fluctuation in seed moisture significantly affects the extent of insect infestation.

iv. Oxygen availability

Oxygen is well known for respiration in living organism. Insects consume 130,000 times more oxygen than that of seeds. Each insect species requires a particular minimum concentration of oxygen for its survival and multiplication and they cannot survive when the oxygen goes below that level so air tight containers are proved to be better for storage of seed.

b. Biotic factors

A number of biotic factors affect the grain quality during storage. They infest and damage the grain up to a great extent.

i. Internal feeder

These insects lay their eggs at pre-harvest stage or during storage of seed and develop inside the seed. The larvae feed inside the endosperm/cotyledon and pupate inside the kernel. To overcome the problem of insect, seed should be cleaned and dried well in sunlight.

ii. External feeder

These insects require broken seeds/powder. They cause damage to the stored seed. The insects feed on embryo and make them totally unfit for germination. The insects and their different stages are generally visible among the seed. To overcome the damage by external feeder seed should be stored in cleaned stage in proper storages and if requires, chemical may also be used.

iii. Mites and fungi

They cause damage especially under high humid tropical conditions. To overcome the attack of mites and fungi cleaned seed should be stored in clean and dried storages.

iv. Rodents

Rodents cause damage to the storages, stored grain and packaging material too. They cause damage up to a great extent and also cause admixture of different lots of seed. To overcome the problem of rodents storages should be in good condition and protection measures should be adopted.

v. Birds

Birds like pigeon and sparrow also damage to the seed, packing material and cause admixture of seed.

SOURCES OF INSECT INFESTATION

The infestation seed starts from the standing crop in the field. The insects lay eggs on the pods and the harvested seeds have the optimum temperature and seed moisture for the development of insects. The infestation is detected at the time of emergence of adults. The sources of infestation are:

a. Field infestation

Bruchids infest the seed crop in the field and come along with harvested produce. The insects multiply and damage the seed.

b. Left over seed of previous year

Left over seed is also a major source of seed infestation. In general it is seen that left over seed carry so many insect which damage the fresh seed too.

c. Un-cleaned processing unit

There are possibilities of presence of insects in the processing units and if the processing units are not cleaned properly the insects present in the processing unit will infest the fresh seed.

d. Old and infested gunny bags

Insects hide in the old gunny bags also infest the seed up to great extent.

e. Infected tractor trolly, trucks and wagons

Vehicles are a major source of infestation if not cleaned properly. Now a day containers are used to transport large quantity of seed and if these are not cleaned or disinfected time to time it will infest the fresh seed too.

f. Insects present in the vicinity

The insects present in the vicinity also infest the seed and damage the quality of seed.

g. Negligence during storage at any time

Adult and larval stages hiding in the cracks and crevices and are major source of infestation of seed. If the storage is not dis-infected time to time they cause heavy damage to the seed.

INSECT-PESTS AND THEIR MANAGEMENT IN STORED SEED

a. Insect-pests

Insect-pest damages the grain and its quality too. In tropical countries, food grains suffers badly from storage insect-pests. Insect-pest infestation leads to poor germination of seed or sometimes totally unfit for sowing. The quality of seed is affected on the basis of damage to the endosperm or embryo. If embryo is damaged, seed fails to germinate and if endosperm is damaged, seed may germinate but the vigour will depend on the extent of damage. Pests hiding in the cracks and crevices of godowns are the major source of infestation. Insects hide in gunny bags, containers and transport vehicles are also one of the major sources of infestation. The peak damage is generally observed from April to September. The main grain pests are as under:

Gram dhora or Bruchid (*Callosbruchus chinensis* Fabricius)

Gram dhora belongs to the family Bruchidae and order Coleoptera. It breeds actively from March to the end of November and hibernates in winter in the larval stage. The eggs are small and oval like glud on grains.

The insect attacks in fields as well as in the store. In field the pest lays eggs on pods and the young larvae bore into grain and complete its developmental stages inside the grain. The larvae damage the grains by feeding inside the grain. The maximum damage is caused from April to September when all the stages of insect are present and reduced in October-November. The larvae feed and breed inside the grain, consuming the entire food content and make them hollow. It is most destructive store grain pest.

a. Control measures

Proper care of grain helps in reducing incidence of bruchid and insurer proper supply of quality seed. Pre-storage and post-storage care, definitely protect the seed from insect-pests and also improve the seed quality.

1. Pre-storage care

i. Implements and equipments used in harvesting, threshing and processing should be clean and insect free.

ii. Disinfect the vehicles before use.

iii. Disinfect the empty seed storage by spraying Melathion (50 EC) @ 0.5 per cent using knapsack sprayer, fill cracks and crevices and leave up to 7 days.

iv. Soak the old gunny bags into boiling water for 15-20 minutes or disinfect old gunny bags by dipping them into 0.2 per cent Melathion (50 EC) emulsion for 10 minutes and dry well in shade.

v. Seed storage in 700 gauge polyethylene bags is equally good as gunny bags.

vi. Store clean and dry seeds with the moisture content 8%.

vii. Dry the seed on black polythene sheet at 40-50°C for 8-10 hours. It helps to overcome the pest attack and have no adverse effect on germination.

viii. Do not store new harvest with carry-over stocks.

ix. Ensure that fresh harvest should not carry any insect infestation.

x. If there is any infestation the seed it must be fumigate before moving to the store.

xi. Store in clean, undamaged and dry stores.

xii. Put the filled gunny begs on wooden boards or straw or bricks to protect the seed from moisture.

xiii. At the time of storage staking should be done up to 80% height only.

xiv. Staking should be done in such a manner that bags don't come in the contact with walls of store.

2. Post-storage care

i. Storage in controlled condition reduces insect infestation effectively. Seed should be stored at low temperature (<20°C) and low relative humidity (<30%).

ii. The moisture content may not be more than 10%, preferably it should be 8%.

iii. Clean the surrounding area of store before storage of seed/grain.

iv. After proper stacking, treat the left over free space by spraying Melathion (50 EC) @ 0.2 per cent.

v. In case of infestation, fumigate the godown with EBD ampul @ 3 ml/100 kg of seed and keep the godown air tight up to 5-7 days to kill all the stages of insect.

vi. To avoid insect damage mixing of Malathion 5 D @ 1 kg/100 kg of seed.

vii. Spray of store with Melation 50 EC @ 0.2% to kill any insect remained in the store.

viii. Use window trap and grain prob to know the presence of insect and adopt control measure as per requirement.

RATS

Rats are major threat to the stored grain and as far as legumes are concerned rats have become serious problem, because chickpea is rich in protein and are palatable in comparison to cereals. In general one rat can eat 20-30 g grain per day and also deteriorate the quality of the seeds by leaving urine and excreta in the seed. Sometimes they cause damage to the standing crop by cutting it and eating seed. They also damage the threshing floor by making holes in to it.

Control measures

1. Damage the holes near seed store regularly.
2. Take care that seed store should be clean.
3. Fill up the holes by cement and concrete.
4. Use rat-catching boxes.
3. Put iron strip on the bottom of wooden gates of seed store.
6. Prepare baits using Zinc phosphide (1 part) + floor (40 parts) + some oil and sugar or Qumarin (25 g) + sugar (20 g) + floor (500 g) + oil (20 g) and put it in godowns or on the path of rats.
7. Use iron bins and well-protected storage.

METHODS OF FUMIGATION

Normally fumigation is done in the room or under cover. The methods of fumigation are:

a. Space fumigation

In this method the entire space of godown is fumigated which helps in complete destruction of insects hiding in the cracks and crevices, besides infesting bagged seed. The quantity of fumigant is used as per volume of the store rather than its occupancy. The small cracks may be seal by plaster of paris, whereas, large cracks or holes should be repaired before fumigation. The effectiveness of fumigation depends on air tight condition of store for a particular time. Outer opening like doors, windows and ventilators may be covered with high density (1000 gauge) polyethylene fumigation sheet and joints may be seal by plastering suitable material or adhesive tape.

b. Fumigation under sheet or cover

It provides better control of insects than space fumigation. Sheet or cover fumigation is generally used in sheds. The stacks of bagged seed are covered by high density (1000 gauge) polyethylene sheet with perfect sealing of joints. The size of sheet should be sufficient to cover the stacks properly with the margin to seal at ground level. The quantity of fumigant is to be calculated as per the quantity of the seed rather than size of the cover. The cover should be protected from the rodents.

c. Fumigation under portable gas tight tent

It is used for *in situ* treatment of material and containers.

REQUIREMENTS OF SAFE STORE

The primary objective of safe storage is to protect the seed from any type of damage may be biotic, abiotic or by natural calamities. To overcome seed damage the storage should have following characteristics:

1. It should be on sufficient height from the ground level.
2. It should be moisture proof.
3. It should allow controlled aeration to limit insect development.
4. It should be capable to make air tight, so that fumigation can be done properly.
5. It should be designed in such a way that cleaning, seed inspection and seed movement can be done smoothly.
6. It should be well protected from rodents, birds and theft.

7. It should be in the close vicinity of the inspectors so that seed may not be damaged in any case.

INSPECTION OF STORAGE

Constant vigilance and frequent inspection of seed/grain store is important to protect the seed from any damage and to detect the infestation. Inspection of seed store including outside areas are essential to judge the insect damage. During inspection the behavior of insect should be kept in mind. Foul odour, cakeing and crushing, flour spots on bags or dockage generally indicate infestation. Regular inspection including all static areas and equipments is essential for good storage. The infestation depends on temperature and moisture content of seed. Therefore, it is better to inspect the storage at an interval of 3-4 weeks time in normal period and fortnightly in monsoon period and coastal areas. If live insect-pest or infestation is detected, fumigated the seed store immediately.

REFERENCES

Avaknavar, J.S. 2003. Management of storage insect-pests in the field and storage. ICAR Sponsored Winter School on "Recent Advances in Hybrid Seed Production Technology in Agriculture and Horticulture Crops" held at Department of Seed Science and Technology, College of Agriculture, UAS Dharwad, Karnataka, 19th Nov. to 9th Dec. 2003. PP: 136-141.

Balasubramani, V. and Prabakar, K. 2008. Insect pests of seeds in storage and their management. Advances in seed science and technology by K. Vanangamudi, N. Natrajan, K. Natrajan, A. Bharthi, R. Umarani and T. Saravanan. Agrobios (India), Jodhpur. PP: 737-743.

Chhabra, K.S.; Lal, S.; Kooner, B.S. and Verma, M.M. 1992. Insect pests of pulses. Identification and control manual. Punjab Agricultural Unive-rsity, Ludhiana and Directorate of Pulses Research, Kanpur.

Indo-Japanese Project on Development of Quality Seed. Modern seed processing and storage. 1999. Indian Agricultural

Research Institute, Regional Station, Karnal-132 001 (Haryana).

Khedar, O.P.; Singh, R.V.; Shrimali, Mahesh and Singh, N.P. 2008. Pulses: Status and Cultivation Technology. Aaviskar Publishers, Distributors, Jaipur (Raj.)

Kooner, B.S.; Taggar, Gaurav K. and Cheeme Harpreet K. 2007. Insect Pests of Pulses-Status and Management. Pulses at a glance. Pulses Section, Department of Plant Breeding, Genetics and Biotechnology, PAU, Ludhiana-141 004.

Kumar, Yogesh; Kumar, Atul; Singh, Pradeep Kumar and Shrivastava, G.P. Improved cultivation of gram in Jharkhand. Birsa Agricultural University, Technology Bulletin 22.

Singh, Shivakant. 2009. How to store pulses properly. Pulse newsletter. July-December 2009. IIPR, Kanpur. 3(2). PP: 7.

Sinha, J.P.; Modi, B.S. and Sinha, S.N. 2001. Post harvest seed handling and quality management in wheat. Indian Agricultural Research Institute, Regional Station, Karnal-132001 (Haryana).

Sinha, S.N. 1999. Safe storage of seeds: Insect pest management. Indian Agricultural Research Institute, Regional Station, Karnal (Haryana).

15

Biotechnology and Molecular Marker Application in Chickpea

S.S. Shekhawat and N.P. Singh

AN OVERVIEW

Biotechnology refers to the process for mass production of useful substance, products for industry, application of various biological organism, medicine and agriculture. Chickpea (*Cicer arietinum* L.) is an important pulse in the tropical and subtropical region of the world which is having rich seed protein content. Chickpea production remains almost stagnant in India even after the release of specific varieties during the last decades. The major reasons for slow progress through conventional breeding method is limited because of less variation and complicated linkage relationships among genes associated with yield traits which requires a period of time in the release of candidate variety. Chickpea suffers yield losses due to abiotic and biotic stresses including climate change which leads to heat stress and cold stress. Limited variation is available in the chickpea against the biotic and abiotic stresses with the introduction of new biotypes and races with high virulent value in specific pest that are the serious threat to meet out genetic solution to the pest and diseases attack. Now-a-days, helicoverpa (pod borer) is the major threat to chickpea cultivation in the present time. Presently large range of pest are there, for which there is no satisfaction level of resistance is available including primary and in special cases secondary gene pool also. Conventional breeding methods plays an important role in maximizing and stabilizing the seed yield of chickpea but this is high time to meet out the maximum variation which is needed for successful chickpea breeding.

In the present scenario cell biology, genetic engineering and molecular markers offer new tools for morphological traits with the help of RFLP, PCR, RAPD, AFLP, ISSR, SNP, DARP and DArTS to identify the molecular markers which are responsible for inheritance of resistance against the biotic and abiotic stress in the candidate variety or genotype or cultivar. Tools used in conjunction with conventional breeding techniques are expected to contribute significantly towards developing chickpea cultivars which are more competitive with their environments and to meet out the consumer preference, value addition and export quality to earn more and more Forex. Rapidly evolving biotechnological approaches can be used effectively for the genetic improvement of chickpea in three ways:

1. Tissue culture and embryo rescues techniques can be used to facilitate introgression of genes from the wild relatives and weedy species.
2. Transformation of desired genes from unrelated species.
3. Molecular markers can be used to tag the useful genes for transfer through marker-assisted selection.

REGENERATION

Grain legumes and chickpea are lacking an efficient regeneration system that is pre requisite for exploitation of direct gene transfer via transformation (Parrott *et al.* 1992). It was initially reported by Mukhopadhyay and Bhojwani

(1978) which indicated regeneration from culture in Chickpea in which induction of organogenesis from various vegetative explants was done. In chickpea shoot regeneration from immature cotyledons (Shri and Davis 1992) and shoot meristem (Brandt and Wakhlu 1993, Dinesh Kumar *et al.* 1994, Chauhan and Singh 2002) has been reported. There are so many limitations of somatic embryogenesis protocols leading to low frequency, inconsistency, genotype specificity and appearance of a callus phase before embryogenesis. Therefore, reproducible high frequency regeneration system may exist from direct organogenesis from the pre-existing meristem. Cotyledonary node and shoot tips has helped in propagating chickpea (Altaf and Ahmed 1986; Chandra *et al.* 1993).

Organogenesis

Tissue culture techniques in breeding programmes is required for highly reproducible regeneration of plant and reliable genotype neutral. Leguminous plants are considered recalcitrant as compared to the members of family Solanaceae and Cruciferae. Direct regeneration of plants from explants without callus phase has been reported from apical meristem tips in chickpea (Bajaj and Dhanju 1979, Kartha *et al.* 1981, Altaf and Ahmad 1986, Rao and Chopra 1989), hypocotyls (Neelam *et al.* 1986), cotyledonary node and cotyledon (Rao and Chopra 1989), embryonic axic (Singh *et al.* 1996) and epicotyls explants (Rao and Chopra 1987).

As both root and shoot grow simultaneously direct regeneration has an edge over indirect morophogenetic pathway. Differentiation of shoots tip cotyledon node and hypocotyls have been reported (Chandra *et al.* 1993, Adkins *et al.* 1995, Rao and Padmaja 1996). Out of three explants, shoot tip and cotyedon node differentiated into multiple shoots with slight callus at the base and bushy roots. Cotyledon node explants differentiated into plants (Table 1). Whole plant regeneration from immature leaflets and callus is obtained by Barna and Wakhlu (1994).

Table 15.1. Present scenario of regeneration in chickpea

Explants	Route	Regeneration	References
Apical meristem	Organogensis	Plantlets	Bajaj and Dhanju (1979), Altaf and Ahmad (1986), Kartha *et al.* (1981)
			Rao and Chopra (1989), Singh *et al.* (1996)
		Direct multiple shoots	Neelem *et al.* (1986)
			Altaf and Ahmad (1986)
			Shiela *et al.* (1991)
			Chandra *et al.* (1993)
			Sharma *et al.* (1998)
			Subhadra *et al.* (1998)
		Multiple shoots with callus phase	Surya Prakash *et al.* (1992)
	Somatic embryogenesis		Chandra *et al.* (1993)
			Sanagawan *et al.* (1989)
			Rao *et al.* (1991)
Hypocotyl	Organogenesis	Multiple shoots	Neelem *et al.* (1986)
			Chauhan and singh (2002)
		Rooting	Adkines *et al.* (1995)
		Multiple shoots with callus phase	Rao and Padmaja (1996)
Cotyledonary node	Organogenesis	Plantlets	Altaf and Ahmad (1986)
			Rao and Chopra (1989
			Chandra *et al.* (1993)
			Sharma *et al.* (1998)

	Somatic Embryogenesis	Multiple shoots	Sanagawan *et al.* (1989)
		Torpedo stage	Shankar and Mohan Ram (1993)
Cotyledons	Organogenesis	Direct regeneration	Rao and Chopra (1989)
		Multiple shoots with callus phase	Surya Prakash *et al.* (1992)
	Somatic Embryogenesis	Multiple shoots	Sanagawan *et al.* (1989)
			Shri and Davis (1992)
			Shankar and Mohan Ram (1993)
		Torpedo stage plants	Sagare *et al.* (1993)
			Sudhavani and Reddy (1994), Adkines *et al.* (1995)
			Eapen and George (1994)
Embryo axes	Organogenesis	Direct regeneration	Singh *et al.* (1997)
		Multiple Shoots	Subhdra *et al.* (1998)
			Sharma *et al.* (1998)
	Somatic Embryogenesis	Plantlets	Chauhan and singh (2002)
Immature	Somatic Embryogenesis	Plantlets	Eapen and George (1994)
Embryonic axes			
Mature embryonic	Somatic Embryogenesis	Plantlets	Suhansini *et al.* (1994)
Axes			
Leaflets	Organogenesis	Direct regeneration	Sagare *et al.* (1993)

	With callus phase	Barna and Wakhlu (1993)
Somatic Embryogenesis	Somatic embryos	Rao and Chopra (1989)
	And regeneration of Plants	Dinesh Kumar *et al.* (1994); Barna and Wakhlu (1993)
	Torpedo stage	Shankar and Mohan Ram (1993)

Somatic Embryogensis

It occurs indirectly through an intervening callus phase or directly from cultured tissues. Induction of somatic embryos was accomplished in the presence of 2,4-D in the medium, while development occurred after its removal of auxins from the medium which result into mimicking of the development events of Zygotic embryogeny. Few reports has been observed of somatic embryogenesis in chickpea from callus Cultures of root segment, shoot disc, cotyledon, hypocotyls, cotyledonary nodes. Shoot apices and immature embryos (Sangawan *et al.* 1989) and Leaflets (Rao and Chopra 1989, Dinesh Kumar *et al.* 1994, Barna and Wakhlu 1993) Sagare *et al.* (1995) obtained somatic embryos from five genotypes of chickpea viz. Phule G 1, Phule G 5, Phule G 12, N 59 and C 235.The regenerated plants were transferred to soil and grown to maturity. Eapen and George (1993) investigated regeneration from somatic embryos using immature cotyledons and immature embryonic auxin.

DISTANT HYBRIDIZATION

Annually Chickpea is susceptible to biotic stresses and suffers heavy yield losses. Biotic factors such as Ascohyta blight, *Fusarium* wilt, and pod borer (*Helicoverpa armigera*) and abiotic stresses such as drought, salinity, water logging, nutrient deficiencies and extremes of temperature contribute significantly to low yield of chickpea. Wild species of chickpea possess genes for tolerance to biotic and abiotic stresses. Wide hybridization technique can be exploited for chickpea variety improvement because wild species of chickpea have resistance to these biotic and abiotic stresses and thus (Table 2). Wild species depicts the characters for good source of resistance to asochyta blight and to botrytis grey mould.

Table 15.2. Ovule culture techniques in Inter-specific hybridization in chickpea using embryo rescue

Cross	Status	References
Cicer arietinum x C. pinnatifidum	18 DAP, ovule were cultured and hybrid shoots were grafted on CL 769 stocks. Hybrid plant with albino shoots produced	Badami et al. (1997)
Cicer arietinum x C. pinnatifidum	18-25 DMP, ovule culture obtained. Formed hybrid which showed albino shoots. This plant resembled male parent	Mallikarjuna (1999)
Cicer arietinum x C. cuneatum	Ovary culture after 4-5 days DAP and hybrid plants were morphologically more similar to C. cuneatum but hybrid plant did not flower	Singh and Singh (1989)
Cicer arietinum x C. echinospermum	20-28, days old embryo excised from ovule & cultured in growth medium, hybrid not formed.	Shiela et al. (1994)
Cicer arietinum x C. judaicum	Successfully hybrids were recovered by embryo-rescue and promising lines were isolated from segregating material	Singh (1997)

C. reticulatum and *C. echinospermum* are readily crossable with *C. arietinum* among wild species. At different centres in India, introgression of genes from these two species is already underway. However, there are numerous cross-incompatibility barriers in inter-specific crosses of chickpea (Singh *et al.* 1999a). Just a few reports are available in chickpea on ovule/embryo culture technique. *Cicer arietinum* was crossed with *C. cuneatum* using ovary/

embryo culture technique and hybrid was produced (Singh and Singh 1989). Singh (1997) reported successful production of inter-specific hybrids involving *C. judaicum*. Recently, inter-specific crosses of *C. arietinum* with *C. pinnatifidum* were reported (Badami *et al.* 1997 and Mallikarjuna 1999). Singh (1997) reported Successful production of inter-specefic hybrids involved C. judaicum.

In vitro selection

Various cells can be selected *in vitro* by their ability to grow in the presence of selection agents such as fungal filtrates or toxins, heavy metal ions, herbicides, high salt levels or amino acid analogues in the culture medium. This method is especially useful for isolating rare variants as large cell populations can be screened but the inheritance of selected trait has not been reported in any crop plant. Callus tolerant to NaCl has been obtained by Gosal and Bajaj (1984) and Pandey and Ganapathy (1984). Gosal and Bajaj (1984) obtained rhizogenesis from NaCI tolerant Callus/ Cell lines. Atrazine resistant callus was initiated from cell suspension culture in chickpea and resistant plants to Atrazine were reported by Rao *et al.* (1991). Callus resistant to Ascochyta blight and subsequently regeneration of plantlets were obtained by Singh *et al.* (1999) Plants resistant to *Ascohyta rabiei* and *Fusarium oxysporum* were reported by surya Prakash *et al.* (1992) (Table 15. 3).

Table 15.3

Target	Status	References
Salinity	Callus	Pandey and Ganapathy (1984)
Salinity	Rhizogensis	Gosal and Bajaj (1984)
Salinity	Plantlets	Jaiswal and Singh (1999)
Fusarium oxysporum and Ascochyta rabiei	Plantlets	Suryaprakash *et al.* (1992)
Ascochyta blight	Callus	Singh *et al.* (1999)
Ascochyta blight	Plantlets	Singh *et al.* (1999)

GENETIC TRANSFORMATION

Genetic transformation has become an important aid to plant physiologists and biochemists in understanding the role of individual genes in the life of plant. There is a major interest in developing transgenic plants with the development of methods for transfer of specific genes and their expression in the new host species since 1983. Chickpea has shown Agrobacterium – mediated and direct gene transfer methods. All the major crops can be infected by wild type of Agrobacterium which produces tumours. So far, most of the transformation experiments in pulse crops are limited to the transfer of marker genes. Only a few reports are available on genetic transformation in chickpea. Transformed callus was obtained using wild strains of Agrobacterium, while roots were obtained by treating chickpea with root inducing strains of Agrobacterium. Transformed chickpea plants were obtained for the first time by Fontana *et al.* (1993) and subsequently by Kar *et al.* (1996) by treating seed derived embryo axes with Agrobacterium tumefaciens strain LBA 4404 harbouring the binary vector pBI 121. The inheritance of the transgenes was not demonstrated in these studies (Table 15.4). Recently, a recombinant gene is transmitting to progeny following A. tumefaciens mediated transformation has been reported in chickpea. (Krishnamurthy *et al.* 2000)

MOLECULAR MARKERS

In recent years, a variety of molecular markers, based on microsatellite or simple repeats (SSRs) have become the markers choice, thus necessitating their development and use in a variety of plant systems. In this review, the basic principles underlying different hybridization based (oligonucliotide fingerprinting and PCR based approaches (STMS, MP-PCR, AMP-PCR/ISSR/ASSR, RAMPs/dRAMPs and SAMPL), making use of microsatellite have been outlined. Different methods for enrichment of genomic libraries for microsatellites have also been outlined. Relevant literature on the subject, giving summary of results obtained using each approach have been reviewed and critically discussed. The review also includes a discussion on literature which deals with the use of microsatellite in genome mapping, gene tagging, DNA fingerprinting, characterization of germ plasm and cytogenetics research. Special emphasis has been laid on the genome of bread wheat, where the work done in the authors laboratory has also been briefly reviewed.

Table 15.4. Current status of genetic transformation in chickpea

Target	Agrobacterium tumefaciens strain	Plasmid	Explants	Marker	Reporter	Status	References
Insect Resistance & herbicide resistance	EHA 101 & C58CI/GV 2260	pIBGUS & p35SGUS INT (Binary vector) pBI 121	Embryo axes	Npt-II	GUS-assay	Analysis of 4 plantlets of TI progeny; none of them GUS +ve; npt II gene detected	Krishnamurthy et al. (2000)
Insect resistance	LBA 4404	pBI 121 (Binary vector)	Embryo axes	Npt-II	GUS-assay	Transgenic Plants transferred to green house.	Kar et al. (1996)

Crop improvement and practical breeding programmes requires Genetic markers. Traditionally, reliance has been made mainly on visible phenotype such as morphological differences between individuals by the breeding programmes. However, markers based on DNA probes have been proved to accurately identify quantitative trait loci (QTLs) for agronomically and physiologically important characters.

The major strength of DNA markers is their potential to reveal almost unlimited number of polymorphisms. Direct applicability in breeding programme is depicted due to features like ubiquitous nature, detection at any developmental stage and independent of environmental effects.

Restriction fragment length polymorphism (RFLPs) is widely used for construction of genetic maps and assessment of genetic variability. Insufficient polymorphism has been found in Isozyme electrophoresis of chickpea which is useful for tagging of wilt resistance (Tullu *et al.* 1998). A larger number of markers can resolved by means of restriction fragment length of polymorphism (RFLPs). Possibilities have been made to generate high density maps of various commercially important crops like rice, tomato, potato, lettuce and soybean. However, a low level of polymorphism (only seven) has been identified and added using this technique to the chickpea gene map (Simon and Muehlbauer 1997). It relies mainly on hybridization of cDNA or low copy genomic probes with complementary unique DNA sequences. On the other hand, use of mini-satellite probes is targeted for hyper variable repetitive sequences. Abundance of micro-satellited in all genomes and its even distribution throughout the genome has proved useful in developing DNA fingerprints of closely related varieties.

A set of markers does not require prior sequence information and use of random combinations of oligonculeotides as primers e.g., DAF. This property has been exploited in microsatellite sites (STMS) approach, and such analyses are very useful for phylogenetic studies. It makes the PCR based microsatellite markers time consuming and expensive, limiting the large scale use of this approach to a few important plants. Highly efficient RAPD markers are considered for discovering polymorphisms which are randomly distributed throughout the genome. The multiple nature of RAPD assay is extremely useful in analysis of molecular taxonomy and population genetics. These markers have been employed widely for mapping genomes (Kim *et al.* 1992) and tagging resistance genes. Two RAPD markers namely CS-27 100 and UBC-170 550 have been reported to be linked with genes imparting resistance to *fusarium* wilt race 1 (Mayer *et al.* 1997) Further, these two markers are located on the same side of resistance locus rather than flanking positions. Analysis of markers has placed these loci in the linkage group VI (Simon and Muehlbaur 1997). These two RAPD Markers previously show to amplify fragments linked to race 4 resistance also amplified fragments associated with race 1 resistance (Tullu *et al.* 1998). RAPD markers were located in 9 map units apart from gene conferring resistance to race 4. Another set of PCR based markers are required with the prior sequence information for designing the primers.

PRESENT OBSTACLES

Recently, researches have provided a better understanding of the molecular biology of stress responses in plants. Identification of genes and promoters has been done which get induced upon exposure of plants to various abiotic stresses like drought, salinity, etc. Consequently, biotechnological tools have been used for stress-tolerance in plants. The potential candidate genes, which are being exploited worldwide for stress-induced proteins such as LEA proteins, RAB proteins, dehydrin, HSPs, Osmotin and annexin, whose primary function is to protect native structure of proteins during stress and/or to repair the denatured proteins. However, there are no such efforts in chickpea to develop transgenic possessing resistance to drought and salinity.

Pod borer, Ascohyta blight and wilt serious are serious biotic constraints in chickpea are. Around 80% to 90% damage is caused by pod borer. Although chemical pesticides are effective, these are cost–intensive and environmentally hazardous. Genes imparting resistance to a variety of insect pests are either not available in the germplasm or extremely difficult to transfer through hybridization. An alternative approach to control pod borer damage is isolation of genes such as Bt. In our country various genes are now being deployed in crops like brinjal, cabbage and cotton to provide resistance against the *Helicoverpa menace*. Besides other insecticidal proteins such as higher plant protease inhibitors, amylase, lecting etc. have also been discovered to impart insect resistance. Non-availability of a highly reproducible, genotype-neutral regeneration and transformation system is the major constraints in developing transgenic chickpea.

Gram is sensitive to fungal diseases such as *fusarium* wilt and Ascochyta blight. Today major challenge for plant breeders is identifying race specific resistance genes and transferring these to adapted backgrounds through conventional breeding. Quicker assessment of susceptibility can be done for development of molecular markers for resistance genes which allows screening of seedlings and reduce the need for maintaining virulent isolates of wilt pathogen. The new class of molecular markers, the micro-satellites has proved to be more useful and highly polymorphic than other markers. The abundance of microsatellites in the chickpea genome is already shown by the hybridization based multilocus fingerprinting techniques using a variety of synthetic oligonucleotide probes of different sequence and length. Thus tagging and pyramiding of race-specific genes into a single genetic background would be helpful and can play significant role in chickpea. The genes imparting resistance to race 1and 4 of *fusarium* wilt have been already tagged using molecular markers. However, there is urgent need to concentrate work on race 2 and 3 of wilt, which also cause considerable damage in India. Such efforts may be able to help in pyramiding genes for multiple races in a single genotype.

RESEARCH PRIORITIES

In our country, biotic and abiotic stresses collectively take a heavy toll of chickpea productivity in agricultural ecosystems and result approximately 45% yield losses annually which amounts to a staggering loss and even if some part of this is saved, it will add substantially to total food basket of the country.

Insect pests and fungal pathogens being important biotic stresses are managed through the use of chemical pesticides and deployment of resistant varieties bred by conventional means besides, indiscriminate use of chemical pesticides leads to adverse effects on environment and human health. Rapid development of genotypes is achieved through the advent of molecular biology and genetic engineering that can tolerate biotic stresses in a more effective manner.

Insecticidal proteins such as higher plant protease inhibitors, amylase and lectins are discovered but the toxicity level of some of these proteins to target pests is not satisfactory and also safety of some of these toxins to human beings is doubtful. Moreover, they have limited utility because their expression leads to only larval growth disturbances but not mortality. Hence, there is an immediate need to use such approach as there are not reliable control measures to tackle the menace of pod borer. But before doing this there is need to develop highly reproducible and genotype neutral regeneration, transformation system.

In our country various genes are being inserted in crops to provide resistance against a range of insect pests and the most common strategy being used is to insert endotoxin gene from *Bacillus thuringiensis* (Bt) into plants. Development of transgenic crop plants is due to refinements in the Bt technology which can express Bt insecticidal proteins or toxins in desired agronomic base. Different strains of Bt produce different insecticidal crystal proteins coded by the cry genes. These are highly toxic to specific insects, mites nematodes, flatworms or protozoan. Among the various insect-specific Bt toxins, some are effective against insect pests specifically *Helicoverpa armigera*. More than 30 plant species have been transformed with a variety of BT genes. After the successful commercialization of Bt transgenic cotton in the USA, maize and potato followed the same path. Bt transgenic crops are cultivated on large scale in the USA, Australia, China, South Africa and Mexico.

Ascohyta blight disease caused by Ascochyta rabiei causes heavy losses (10-100%) especially in epiphytotic years when conditions of high relative humidity (90-95%) and temperature (20-25°C) prevail during the growing season. Chemical control to Ascochya blight has also been found ineffective in those conditions. Besides, chickpea varieties earlier released as blight resistant have now become susceptible due to change in the virulence of the pathogen. Conventional breeding methods have not been successful in developing blight resistant variety due to unavailability of suitable donors in primary gene pool. The tissue culture techniques like embryo rescue, induction of soma-clones and *In Vitro* selection of cell lines can be used as an alternative tool to avoid yield losses due to blight disease.

The morphological markers being used in the past have undesirable effects on plant phenotype and hence, their use in chickpea improvement has been limited. The tools of molecular biology have provided an opportunity to develop large number of phenotypically neutral genetic markers, which can reveal sufficient level of polymorphism at DNA level, which unfortunately could not be obtained by RAPD and RFLP. However, the use of microsatellite and AFLP markers has shown considerable polymorphism in chickpea. There is an urgent need to develop molecular markers for resistance genes as it could lead to quicker assessment of susceptibility, allow screening of seedling and reduce the need for maintaining virulent isolates of wilt pathogens. Tagging and pyramiding genes into a single genetic background would be helpful and can play significant role in developing resistance against disease in chickpea. Resistance genes to race 1 and 4 of *fusarium* wilt has already been tagged using molecular markers. Efforts are required to concentrate work on race 2 and 3 wilt, which also cause considerable damage in India.

UPCOMING THRUST

In future, chickpea improvement will be based principally on recombination breeding programmes supplemented by advanced techniques for generation genetic variation and selecting desirable genotypes. The main aim of the non-conventional strategies will be to replace empirical type-based aspects of recombination breeding with more directed and predictable intensive analysis of agronomic traits including the cloning of key genes and the development of phenotype-independent marker-based selection indices for traits where such genes have not been identified. However, the cost benefit ratio and availability of resources for the more intensive biotechnological tools will ultimately determine their use.

REFERENCES

Adkins AL, Godwin ID and Adkins SW 1995. An efficient *in vitro* regeneration systems for Australian grown chickpea (*C. arietinum* L.) *Australian Journal of Botany* 43.(5) 491-467

Altaf N and Ahmad MS 1986. Plant regeneration and propagation in chickpea (*Cicer arietinum* L.) through tissue culture techniques. Pp 407-417. (in) Proceedings, Symposium Nuclear Techniuques and *in vitro* culture for plant Improvement (19-23 August 1985), FAO, IAEA, Vienna, Austria.

Badami PS, Mallikarjuna Nand Moss JP 1997. Interspecific hybridization between *Cicer arietinum* x *C. pinnatifidum. Plant Breeding* 116: 393-395.

Bajaj YPS and Dhanju MS 1979. Regeneration of plants from apical meristem tips of some legumes. *Current Science* 48: 906-907.

Barna KS and Wakhlu AK 1993. Somatic embryogenesis and plant regeneration from callus culture of chickpea. *Plant Cell Report* 12: 521-524.

Barna KS and Wakhlu AK 1994. Whole plant regeneration in *Cicer arietinum* from callus culture via organogenesis. *Plant Cell Report* 13: 510-513.

Brandt EB and Hess D. 1994. *In vitro* regeneration and propagation in chickpea from meristem tips and cotyledon nodes. *In vitro* Cell Development. *Biology* 30: 75.

Chandra R, Chatrath A, Raghuveer, Polisetty R and Khetrapal S. 1993. Differentiation of *in vitro* grown explants of chickpea. *Indian Journal of Plant Physiology* 36: 121-124.

Chauhan R and singh NP. 2002 Plant regeneration via somatic embryogenesis in chickpea (*Cicer arietinum* L.) *Indian Journal of Genetics & Plant Breeding* 62 (4): 319-321

Dinesh Kumar V, Kirti PB, Sachan JKS and Chopra VL 1994. Plant regeneration via somatic embryogenesis in chickpea (*Cicer arietinum* L.) *Plant Science* 109: 207-213

Eapen S and George L 1994. Somatic embryogenesis in *C. arietinum* L: influence of genotypes and auxins. *Biologia Plantarum* 36: 343-349

Eapen S and George L. 1993. Somatic embryogenesis in peanut, influence of growth regulators and sugars. *Plant Cell Tissue & Organ Culture* 35: 151-156

Fontana GS, Santini L, Caretto S, Frugis G and Mariotti D. 1993. Genetic transformation in the grain legume *Cicer arietinum* L (Chickpea) *Plant Cell Reports* 12: 194-198

Gosal, SS and Bajaj YPS. 1984. Isolation of NaCl resistant cell lines in some grain legumes. *Indian Journal of Experimental Biology* 22: 209-214

Gupta, P.K. & R.K. Varshney. The development and use of microsatellite markers for genetic analysis and plant breeding with emphasis on bread wheat.

Jaiswal R and singh NP 1999. Differential response of chickpea genotypes against NaCI under *in vitro* culture. Pp 50. (in) proceedings, National Conference on Plant Biotechnology: Towards strategic Agriculture and Drug Development (15-17 March 1999), Lucknow, India

Kar S, Basu D, Das, S, Ramkrishnan NA, Mukherjee, P, Nayak P and Sen SK. 1997. Expression of cry IA (C) gene of *Bacillus thuringiensis* in transgenic chickpea plants inhibits development of pod borer (*Heliothis ormigera*) larvae. *Thransgenic Research* 6: 177-185.

Kar S, Johson TM, Nayak P and Sen S K. 1996. Efficient transgenic plant regeneration through Agrobacterium –mediated transformation of chickpea (*Cicer arietinum* L.) *Plant Cell Reports* 16: 32-37

Kartha KK Pahl K, Leung NL and Mroginski LA. 1981. Plant regeneration from meristems of grain legumes: soybean, cowpea, peanut, chickpea and bean. *Canadian Journal of Botany* 59: 1671-1689

Kim DH, Martyn RD and Magill CW 1992. Restriction fragment length polymorphism groups and physical map of mitochondrial DNA from *Fusarium oxysoprum* f. sp. *niveum*. *Phytopathology* 82: 346-352.

Krishnamurthy KV, Suhasini K, Sagare AP, Meixner M, de Kateh A, Pickardt T and Schieder 2000. Agrobacterium- mediated transformation of chickpea (*Cicer arietinum* L.) embryo axes. *Plant Cell Reports* 19: 235-240.

Mallikarjuna N. 1999, Ovule and embryo culture to obtain hybrids from interspecific incompatibility pollinations in chickpea. *Euphytica* 110: 1-6.

Mayer MS, Tullu A, Simon CJ, Kumar J, Kaiser WT, Kraft JM and Muehnbauer FJ. 1997. Developments of a DNA marker for *Fusarium* wilt resistance in chickpea. *Crop Science* 37: 1625-1629.

Mukhopadhyay A and Bhojwani SS. 1978. Shoot bud differentiation in tissue cultures of leguminous plants. *Journal of Plant Physiology* 88: 263-268.

Neelam A, Reddy CS and Reddy GM. 1986. Plantlet regeneration from callus cultures of *Cicer arietinum* L. *International Chickpea Newsletter* 14: 12-13

Pandey R and Ganapathy PS. 1984. Effect of NaCI stress on callus culture of C.arietinum L. cv, BG-203: Growth and Ion accumulation. *Journal of Experimental Biology* 35: 1194-1199.

Parrott WA, Bailey MA, Durham RE and Mathews HV. 1992. Tissue culture and regeneration in legumes. Pp 115-148. (in) Biotechnology and Crop Improvement in Asia (Ed. J.P. Moss), ICRISAT, Patancheru, India.

Rao BG and Chopra VL. 1987. The influence of media on callusing and organogenesis in chickpea (*Cicer arietinum* L.) *International Chickpea Newsletter* 17: 7-10.

Rao BG and Chopra VL. 1989. Regeneration from apical meristem, stem nodes and cotyledon of chickpea (*Cicer arietinum* L.) *Indian Jounal of Pulses Research* 2(1): 20-24.

Rao S and Padmaja S. 1996. Genotypic and explants differences in establishment of callus and shoot regeneration in chickpea (*Cicer arietinum* L.). In: Role of Biotechnology in Pulse Crops (eds. Jafar Nizam, Irfangvy Khan and S. A. Farook. pp 159-170.

Rao S, Basavaraj K and Madhava Naidum .1991. Isolation of atrazine resistant cell lines and regeneration of plantlets in *Cicer arietinum* (L.) Chickpea : herbicide resistance generated by somaclonal variation. Pp 478-479. *In:* Conference Abstract. International Symposium, 1991, 11th Meet, Long Ashton.

Riazuddin S and Husain T. 1993. Transformation in chickpea (*Cicer arietinum* L.). pp 183-193. *In:* Biotechnology in Agriculture and forestry, Volume 23. Plant protoplasts and Genetic Engineering IV (ed. Y.P.S. Bajaj), Springer-Verlag, Berlin, Germany.

Sagare AP, suhansini K and Krishnamurthy KV. 1993. Plant regeneration via somatic embryogenesis in chickpea (*Cicer arietinum* L.) *Plant Cell Reports* 12 : 652-655.

Sagare AP, Suhasini K and Krishnamurthy KV 1995. Histology of somatic embryo initiation and development in chickpea (*Cicer arietinum* L.) *Plant Science* 109: 87-93.

Sangawan V, Choudhary VK, Sareen PK, Yadav N and Chaudhary JB. 1989. Regeneration from Callus cultures of chickpea genotypes. *International Chickpea Newsletter* 21: 11-13

Shanker S and Mohan Ram Hy. 1993. Induction of somatic embryogenesis in the tissue cultures of chickpea (*Cicer arietinum* L.) *Proceeding of Indian National Science Academy Part B. Biological Science* 59: 629-632

Sharma L and Amla DV. 1998, Direct Shoot regeneration in chickpea (*Cicer arietinum* L.) *Indian Journal of Experimental Biology* 36: 605-609.

Shiela VK, Moss JP, Gowda CLL and van Rheenan HA, 1991. A protocol for micropogation of Cicer Species. *International Chickpea Newsletter* 24: 28-30.

Shiela, VK, Moss JP Gowda CLL and Van Rheenan HA. 1994. Interspecific hybridization between *Cicer arietinum* and wild Cicer species. *International Chickpea Newsletter* 27: 11-13

Shri PV and Davis TM 1992. Zeatin induced shoot regeneration from immature chickpea cotyledons. *Plant Cell Tissue & Organ Culture* 28: 45-51

Simon CJ and Muehlbauel FJ 1997. Construction of a chickpea linkage map and its comparison with maps of pea and lentil. *Journal of Heredity* 88: 115-119.

Singh A .Singh NP Asthana AN and Gurha SN 1999. *In Vitro* selection against Ascorhyta blight of chickpea (*Cicer arietinum* L.) *Journal of Biochemistry & Biotechnology* 8: 110-119

Singh A, Singh NP and Asthana AN. 1997. Callus induction and direct regeneration from immature embryo in chickpea (*Cicer arietinum* L.) *International Chickpea & Pigeonpea Newsletter* 4: 39-40.

Singh A. 1997. Genetical and tissue culture studies in chickpea. Ph.D. Thesis, Kanpur University, Kanpur, India.

Singh NP and Singh BD. 2003. Biotechnological research In Chickpea. *Chickpea Research in India*. **PP** 289-301

Singh NP, Singh A and Asthana AN 1999a. Studies on interspecific crossability barriers in Chickpea, *Indian Journal of Pulses Research* 12: 13-19

Singh RP and Singh BD. 1989. Recovery of rare interspecific hybrids of genus *Cicer arietinum* and *C. cuneatum* through tissue culture. *Current Science* 58: 874-876

Subhadra, Vashisht RK, Chowdhury B, Singh M and Sareen PK 1998. Multiple shoots from cotyledonary node explants of non-nodulating genotype (ICC435M) of chickpea. *Cicer arietinum* L. *Indian Journal of Experimental Biology* 36: 1276-1279

Suhasini K, Sagare AP and Krishnamurthy KV. 1994. Direct Somatic embryognesis from mature embryo axes in chickpea (*Cicer arietinum* L.) *Plant Science* 102: 189-194.

Surya Prakash, Yadav RC, Chowdhury JB and Sareen PK. 1992. Transformation of *Cicer arietinum* L. with Agrobacterium rhizogenes. *International Chickpea Newsletter* 26: 13-14.

Tullu A, Muehlbauer FJ, Simon CJ Mayer NS Kumar J, Kaiser WJ and Kraft JM 1998. Inheritance and linkage of a gene for resistance to race 4 of Fusarium wilt and RAPD markers in Chickpea. *Euphytica* 102: 227-232.

GLOSSARY

Abnormal growth- Deviation from normal growth due to various biotic and abiotic factors like disease, insects-pests, soil condition, humidity, temperature etc.

Adaptability- The capability of adjusting in a new eco-system to cope with changing condition.

Acidic Soils- The soils with pH less than 6.5 and which respond to liming may be considered as acidic soils.

Alkaline Soils- Alkali soils are formed due to concentration of exchangeable sodium and high pH. Because of high alkalinity resulting from sodium carbonate the surface soil is discoloured to black; hence the term black alkali is used.

Amino Acid- A molecule which is the building block of protein. They contain basic atoms of life like nitrogen carbon, hydrogen and oxygen. All organisms contain at least 20 different types of amino acids. Some bacteria are able to make all amino acids, unlike humans which cannot make about 8 amino acids for which they must rely on other organisms, like microbes and plants.

Basic/Nucleus seed- It is I-category in seed multiplication chain produced by concerned breeder/institute. It is purest form of seed and produced in close watch of breeder and is source of breeder seed.

Breeder seed- It is II category in seed multiplication chain and is progeny of nucleus seed. It is produced by concerned or the sponsored breeder/institution under proper isolation distance.

It is the source of foundation seed. Its germination should be 100% and containers shall carry golden yellow tags.

Brittle- Easily breakable.

Catterpillar/larvae- The immature stage (larvae) between egg and pupa of a butterfly, moth or sawfly.

Cereal- Cultivated plant which produces grain used to feed human and domestic animals. Ex. Barley, wheat, oat, Bajara etc.

Certified seed- This is IV category in seed multiplication chain and progeny of breeder or foundation seed. It is the seed that is certified by the seed certification agency established under section 8 of Seeds Act, 1966. Similar to foundation seed, certified seed is also certified by the state seed corporation agency. This seed is used for commercial production. The containers shall carry white tags.

Character- A distinctive but not necessarily invariable feature exhibited by all individuals of a group and capable of being described or measured; e.g. colour, size, performance etc.

Chemical control- Any management practice which rely on chemicals.

Chromosome- Thread like structure in the nucleus carries genes. Each species has a characteristic number of chromosomes.

Container- A box, bottle, basket, tin, barrel, case, bag, wrapper or other thing in which any article or thing is placed or packed.

Contaminant- Any foreign component present in another substance.

Crop diversification- Growing of a variety of crops with a view to make a shift from the prevailing system.

Cropping systems- Cropping pattern used on a farm and their interaction with farm resources, other farm enterprises and available technology.

Cross pollinated crop- The crops in which pollens are transferred from anther to stigma of the other plant.

Dal- Pulses when decorticated and split are known as dal.

Disease- Abnormal condition of a plant due to influence of a pathogen.

Dominant- A phenotype that is expressed in organisms either homozygous or heterozygous for the corresponding allele.

Downey growth- A plant disease in which spores of fungus appear as a downey growth on the lower surface of leaves, stem and fruits.

Dormant seed- A viable seed which does not germinate for a certain period even under favourable conditions is known as dormant seed.

Drainage- Removal of water from land that is too wet by using drains or ditches.

Drought- A prolonged abnormal dry weather.

Drought resistant- Ability of a plant to stay alive, grow and ultimately produce grain under water stress condition.

Fertile- 1. Land that has enough nutrients to support abundant plant growth. 2. Plants that are reproductive and can produce fruits, seeds, spores or pollen grains.

Fertility- 1. Ability to produce viable offspring. 2. The status of the soil with respect to its ability to supply nutrients essential for plant growth.

Field inspection- An official inspection of seed fields conducted by officials of Certification Agency or their authorized representative.

Friable- A term applied to soils that when either wet or dry crumble easily between the fingers.

Erratic- Un-predictable behavior.

Fodder crop - Crop grown for feeding the animals.

Forage - Any kind plant material used for grazing or feeding the animals.

Foliar Application - This refers to the spraying on leaves of growing plants with suitable fertilizer solutions. These solutions may be prepared in a low concentration to supply any one plant nutrient or a combination of nutrients.

Foundation seed - This is III category in seed multiplication chain. The breeder seed procured by different public or private seed companies is multiplied in the form of foundation seed by NSC, SSC, SFCI, SAU's and other public/private companies at their farms or identified growers. The seed production plots are dully certified by the concerned state seed certification agency.

Fumigation - Application of pesticide in gaseous or volatile liquid form, usually made in an enclosed area or incorporated in the soil.

Fungicide - Any substance that kills or inhibits the growth of a fungus.

Fungus - An organism with no chlorophyll, reproducing by sexual or asexual spores, usually having mycelia with well marked nuclei.

Generation - One complete life cycle.

Genetic purity - An individual genetically true to type.

Genetic sterility - Male sterility conditioned by nuclear genes, transmitted by either male or female parent.

Genotype - Genetic constitution of an individual or group sharing a specific genetic makeup.

Germination - The process of initiating growth in seed, spores or plant propagules.

Grain - A small, dry, one seeded fruit that does not open at maturity.

Growing season - Period of year during which crop grown and matures.

Grow-out test - test performed to determine the genuineness of seed as species or variety.

Hard seed - Seeds having a seed coat impervious to water or oxygen required for germination.

Hectare - 10,000 sq. meters area.

Herbicide - Chemical that kills or retards plant growth.

Host - An organism that provides food or shelter for another organism.

Immature seed - Seed with still greenish caryopsis.

Immune - Cannot be infected by a given pathogen.

Infection - Process of penetration of host by a pathogen.

Inert matter- Non-biological products present in biological matter.

Insecticide- Chemicals used to kills insects.

Inseparable - Difficult to separate.

Inter-cropping- Two crops of different height, canopy, adaptation, root system and growth habit grown simultaneously in such a way that they accommodate each other with least competition.

IPM- IPM is a system that in the context of the associated environment and the population dynamics of the pest species utilizes all suitable techniques and methods in as compatible manner as possible and maintains the pest populations at levels below those causing economic injury. In integrated pest management both crop and pest are seen as part of a dynamic agro-ecosystem.

Irrigation- Application of water to the soil for crop production.

Isolation- It is an act of keeping the seed plot away from the sources of contamination.

Isolation- Keeping the seed crop away from the sources of physical and genetical contamination.

Isolation distance- The minimum distance that has to be maintained between the seed crop and contaminant to maintain its purity.

Kind- One or more related species or sub-species of crop plants each individually or collectively known by one name such as mungbean, urdbean, groundnut etc.

Larva- The immature stage of an insect between egg and pupa.

Legume- Legumes are herbaceous woody plants that produce seeds in pods. They have a symbiotic association with root-colonizing bacteria called *rhizobia* and are rich in protein. Examples of legumes are peas, beans, alfalfa, mungbean, chickpea, pigeonpea, groundnut and clusterbean etc.

Local Varieties- Local varieties are nothing but traditional varieties but which are susceptible to diseases and pests and are having long duration for maturity.

Localised placement- This method refers to the application of fertilizers into the soil close to the seed or plant. Localised placement is usually employed when relatively small quantities of fertilizers are to be applied. Localised placement reduces fixation of phosphorus and potassium.

Management- Art of managing the things.

Manures- Manures are organic or inorganic substances applied to the soil to supply one or more nutrients to plants to obtain increased yields.

Mites- Minute invertebrate belongs to the phylum Arthropods, class Arachnida.

Mixed-cropping- Two or more than two crops grown in mixed stand.

Monitoring- Observing and noting specific details. In seed monitoring, monitoring is done to see and maintain its genetical purity.

Moisture - Amount of water present.

Mosaic - Disease characterized by non uniform coloration with light green and yellow patches, usually caused by virus.

Moth- An adult of insect order Lepidoptera, sub order Heterodera.

Mottling- Patches or spots of different colors.

Mutulistic- Both organisms benefit from the association.

Nematode- Thread like round worms cause disease through rotting of plant roots.

Nematocide- Chemical used to kill nematodes.

Necrosis- Localized death of tissues usually characterized by browning and desiccation.

Nitrogen fixation- *Rhizobia* converts atmospheric nitrogen in into a form of nitrogen that the plants can use, this conversion of nitrogen is called "nitrogen fixation."

Nodule -A special structure of roots of certain plants which contain the symbiotic nitrogen fixing bacteria called *Rhizobium*.

Nutrients- The necessary components of food that microbes and all life require to exist and grow. Essential elemental nutrients include such basic molecules as nitrogen, carbon, oxygen, hydrogen, phosphorus, potassium and trace minerals.

Objectionable weed- Plants of weed whose seeds are difficult to separate once mixed with the crop seed and which has smoothening effect on seed crop, difficult to eradicate once established, has a high multiplication ratio thus multiply fast and may serve as an alternate host for pest and diseases.

Off-type/rogue plant- Plant does not confirm to the varietal characteristics described by the breeder.

Organic fertilizers- Organic fertilizers include both plant and animal bi-products. They are slow acting. Organic nitrogen fertilizers include oil cakes, fish manure, dried blood from slaughter houses etc., where as organic phosphorus from bone meal and organic potassium from cattle dung ash, wood ash, leaf mould, tobacco stems and water hyacinth.

Organic Matter- The organic residues left behind after decay of plants and animals. The main source of soil organic matter is plant tissue. Animals are subsidiary source of soil organic matter.

Packaging- It is the process of filling, weighing and sewing of bags.

Pallet- Platform.

Pest- Pest is an organism that causes damage resulting in economic loss to a plant or animal. It can also be said that pest is a living organism that thrives at the expense of other living organism.

Physical purity of seed- Seed free from inert matter, including broken seed.

Plant Growth Promoting *Rhizobacteria* (PGPR) - Root colonizing bacte-ria (*Rhizobacteria*) that exert beneficial effects on plant development via direct or indirect mechanisms have been defined as plant growth promoting *rhizobacteria*.

Pre-emergence- The period after sowing and before germination of seed.

Previous crop- Crop grown in immediate previous season.

Processing- Preparation of harvested seed for marketing. It includes hand-ling, shelling, pre-conditioning, dry cleaning, size grading, upgrading, treating and packaging.

Produce- Something which is produced and harvested such as seed, grain etc.

Production- Process of transformation of certain resources or inputs (land, fertilizer, labour, seed, pesticide etc.) into products or output.

Productive soil- A soil in which the chemical, physical and biological conditions are very favorable for the production of the crops.

Productivity- The amount of output per unit of input expressed in physical or value terms.

Producer- A person or agency grow and distribute seed in accordance with the procedures and standards of Certification Agency.

Protein- An important molecule that is found in all life. Proteins are polymers that are long strings of individual amino acids.

Pulses- Cultivated legumes which produce pulse, used to feed human beings and sometimes domestic animals. Eg. chickpea, pigeonpea, lentil, mungbean, mothbean etc.

Pupa- Non feeding, inactive stage between larva and adult in insects.

Pure seed- Seed that is genetically true to type.

Quality- Distinguishing character of the degree of excellence which is possessed.

Quality seed- Quality seed can be defined as seed of an improved variety which is high in species, varietal and physical purity; has high germination and vigour; is free from weeds and seed borne disease; has low moisture content and is uniform.

Reproductive phase- A period of crop growth from penicle initiation to panicle maturation.

Resistance- The ability of an organism to withstand completely or in some degree against the attack of a pathogen or insect-pest.

Resistant- Possessing the ability to withstand completely or in some degree against the attack of a pathogen or insect-pest.

***Rhizobium*-**Nitrogen fixing bacteria that lives in soil and forms a symbiotic association with the root cells of leguminous plants.

Rodents- Rats, mice and guinea pigs.

Rogue- Rogues are undesirable plant such as weeds, off-types, other variety plants, plant affected with diseases, pollen shadders etc.

Roguing- Systematic removal of undesirable plants from the seed crop to maintain genetic purity of a variety at field level.

Root- The descending axis of the plant, usually below ground, serving to anchor the plant and absorb water and minerals from soil.

Rot- The softening, discoloration and often disintegration of succulent plant tissue as a result of fungal or bacterial infection.

Row- The lines along which a number of plants are planted or arranged.

Runoff- A portion of rainfall or irrigation water draining from the fields.

Saline Soils- The saline soils contains toxic concentration of soluble salts in the root zone. Soluble salt consists of chlorides and sulphates of sodium, calcium, magnesium. Because of the white encrustation formed due to salts, the saline soils are also called white alkali soils.

Seed- A fertilized and ripened ovule that contains a viable embryo, a store of food and a protective covering.

Seed is propagule or dia-spore or any planting material, which can be multiplied sexually (true seed) or asexually (vegetative plant parts such as roots, tubers, stem, branches, leaves, bulb and rhizomes) to produce its own type. or

Sexually or vegetatively propagated planting materials which are used for seeding and planting and as such should be free from any infection (pests and diseases) and should give a good crop stand by good seeding.

Seed borne disease– Disease that is carried within (infection), on (infested), or with (admixture/contaminated) the seed.

Seed certification- A scientific and systematically designed process to secure, maintain, multiply and make available seeds and vegetative propagating material of superior kind and varieties to ensure:

1. General identity w.r.t. distinctness, uniformity and stability (DUS).
2. High degree of physical purity.
3. Physiologically viability to produce vigorous seedlings.
4. Free from all seed-borne diseases, objectionable weed seed & other crop seeds.

Seed Dormancy-Seed dormancy refers to the resting stage of embryo with low germinability of viable and freshly harvested grains. It is also defined as inability or failure of perfectly matured seed to germinate even when placed under conditions favourable for germination.

Seed Moisture- The formation of germination inhibitors was accelerated under reduced oxygen tension and thereby the water content of the seed is increased which deteriorates viability.

Seed-pests- Insects that directly attack on the reproductive parts of the plants and decrease the quality and production.

Seed processing- Refers to all steps involved in the preparation of harvested seed for marketing. It includes handling, shelling, pre-conditioning, dryclining, grading, treating and packaging.

Seed replacement rate- Per cent of acreage sown by improved seed.

Seed treatment- Treatment of seed by biological, chemical or a physical agent, usually to protect the seed and plants from diseases and insect-pests or to stimulate germination or plant growth. or

Seed treatment refers to the application of fungicide, insecticide or a combination of both to seeds, such as to disinfect and disinfest them from seed-borne or soil-borne pathogenic organisms and crop pests both in field and in storage. It also refers to the subjecting of seeds to solar energy exposure, immersion in conditioned water etc.

Seed Viability-Seed viability is defined as "The capability of a seed to show living properties like germination and growth".

Selfing- Self-pollination, artificially or naturally.

Self-pollinated crop- The crops in which pollens are transferred from anther to stigma of the same plant.

Self-pollination- Transfer of pollen from anther to stigma of the same flower or other flower of same plant.

Soil erosion- The removal of top fertile soil by water, wind and gravity.

Soil fertility- Soil fertility deals with the nutrient status or ability of soil to supply nutrients for plant growth under favourable environmental conditions such as light, temperature and physical conditions of soil.

Soil fertility- Soil fertility may be defined as the inherent capacity of soil to supply plant nutrients in adequate amount and in suitable proportion and free from toxic substances.

Soil fumigation- A method of soil treatment using chemicals to disinfect the soil from diseases, nematodes and insects.

Soil organisms- The organisms living in the soil. Soil is the habitat for enormous number of living organisms. Some of these organisms are visible to naked eye whereas, others can be seen by microscope only.

Soil health- The capacity of soil to function within ecosystem boundaries to sustain biological productivity, maintain environmental quality and promote plant and animal health.

Soil inorganic matter- The inorganic constituents of the soil, comprising carbonates, soluble salts, free oxides of iron, aluminium and silica in addition to some amorphous silicates.

Soil pH- The negative logarithm of hydrogen ion (H^+) concentration is called pH. Soil pH may be acidic, basic or neutral.

Soil Productivity- Soil productivity is defined as the capability of the soil for producing a specified quantity of plant produce per unit area and the ability to produce sequence of crops under a specified system of management.

Soil structure– The arrangement of primary soil particles into compound particles. It is one of the important properties of soil, since it influences aeration, permeability and water capacity. The principal forms of soil stricter are; platy, prismatic, columnar, blocky and granular.

Soil texture- The relative proportion of sand, silt and clay particles in a mass of soil. The principle textural classes are; sand, loamy sand, sandy loam, loam, silt loam, silt, sandy clay loam, clay loam, silty clay loam, sandy clay, silty clay and clay. The sand, loamy sand and sandy loam classes may further divided into course, fine and very fine.

Species- Group of closely related individuals.

Sterile- Unable to reproduce sexually.

Sterility- Unability to produce viable seeds.

Strain- A group of similar individuals within a variety.

Sugar- A sweet tasting substance produced by many organisms which is used as an energy storage molecule.

Symbiosis- Symbiosis is an ecological relationship between two or more different species living together in direct contact. There are several types of symbiotic relationships, including parasitism, in which one organism hurts the other; commensalism, in which one organism benefits without hurting the other and mutualism, in which both partners gain from the association. In microbial ecology symbiosis refers to mutually beneficial relationships.

Tolerant- Ability to withstand when attack by biotic or abiotic agent without suffering serious injury.

Vapour-proof container- Container does not permit entrance or exit of moisture.

Variety– A group of closely related plants of common origin and similar characteristics within a species.

Vector- Any organism which transmits disease from diseased plant to healthy one.

Viable- Alive, able to live, grow and develop.

Viable seed- Seed which germinate and develop into a plant.

Vigour- Relative robustness of an individual in comparison to other individuals of the same species.

Virus- An infectious sub microscopic particle causes diseases in plants and animals.

Volunteer plant- Plants of crop species same as that of the seed crop comes from the self seeding or previous season crop.

Weed- Any unwanted plant in the field or any plant that grows where it is not wanted, having negative value within a given management.

Weedicide- Any chemical that kills the weeds and do not affect the main crop.

Weeding- To remove competing vegetation around the main crop or trees.

Well drained- The excess water may be drained easily.

Wilting- The loss of turgidity in plant tissue where the intake of water is insufficient to replace the loss by transpiration or other means, causing a deflation of the plant cell.